IRONMAKING

THE HISTORY AND ARCHAEOLOGY
OF THE IRON INDUSTRY

IRONMAKING

THE HISTORY AND ARCHAEOLOGY OF THE IRON INDUSTRY

Richard Hayman

TEMPUS

WIRRAL LIBRARY SERVICES	
500014925341	
Bertrams	29.03.07
338.4766914	£18.99
HAY	WC

First published 2005

Tempus Publishing Limited
The Mill, Brimscombe Port,
Stroud, Gloucestershire, GL5 2QG
www.tempus-publishing.com

© Richard Hayman, 2005

The right of Richard Hayman to be identified as the Author
of this work has been asserted in accordance with the
Copyrights, Designs and Patents Act 1988.

All rights reserved. No part of this book may be reprinted
or reproduced or utilised in any form or by any electronic,
mechanical or other means, now known or hereafter invented,
including photocopying and recording, or in any information
storage or retrieval system, without the permission in writing
from the Publishers.

British Library Cataloguing in Publication Data.
A catalogue record for this book is available from the British Library.

ISBN 0 7524 3374 1

Typesetting and origination by Tempus Publishing Limited
Printed in Great Britain

CONTENTS

	Acknowledgements	6
1	The 'popular and universal metal'	7
2	The bloomery	11
3	The blast furnace and finery forge	19
4	The coke iron industry	34
5	Puddling	46
6	The nineteenth century	64
7	Cast iron and engineering	86
8	Ironmasters	94
9	Iron workmen	104
10	Steel	116
11	Archaeology and conservation	125
	Select bibliography	150
	Index	157

ACKNOWLEDGEMENTS

This book draws upon several years of work on the iron industry and I would like to acknowledge the influence and support of many people over that period, especially Dr Barrie Trinder, Wendy Horton and staff of the Royal Commission on the Ancient and Historical Monuments of Wales (RCAHMW), who kindly allowed photographs from upland survey work to be reproduced here. I would also like to thank John Powell, librarian at the Ironbridge Gorge Museum Trust, for his help in compiling illustrations for this book.

1

THE 'POPULAR AND UNIVERSAL METAL'

Ironmaking was fundamental to the industrial revolution. 'The second manufacture in the kingdom' is how the ironmaster John Crowley (1689–1728) described the industry in 1717, acknowledging only the woollen industry as making a greater contribution to the nation's wealth. By the end of the century iron was an important and expanding industry in most of Britain's coalfields and epitomised the new technological society of the industrial revolution. It encompassed the smelting and refining of raw material, and the manufacture of a wide range of finished products – large items such as bridges and armaments, and smaller mundane things like nails, horseshoes and steel cutlery. According to the Shropshire ironmaster William Reynolds (1758–1803), in iron ore 'lay coiled up a thousand conveniences of mankind . . . the steam engines, the tramways, the popular and universal metal that in peace and war should keep pace with, and contribute to, the highest triumphs of the world'. Ironmasters were not slow to point out that iron was vital to the prosperity of the kingdom, and that their business was an act of patriotism. Cannon from British foundries pounded the nation's enemies from Blenheim to Waterloo and protected its maritime trade. Moving forward, it was British iron rails that conquered the world in the nineteenth century (although, with *chemin de fer*, it was the French who coined the best term).

Iron manufacture epitomised the triumph of man over nature, in harnessing its elemental powers. By the end of the eighteenth century places like Coalbrookdale in Shropshire and Merthyr Tydfil in upland Glamorgan enjoyed the status of tourist attraction. The strangeness and drama of an ironworks was captured by many contemporary authors. Visiting Merthyr Tydfil in 1848, Charles Cliffe wrote:

> The scene is strange and impressive in broad daylight, but when viewed at night is wild beyond conception. Darkness is palpable. The mind aids reality – gives vastness and sublimity to a picture lighted up by a thousand fires. The vivid glow and roaring of the blast furnaces near at hand – the lurid light of distant works – the clanking of hammers and rolling mills, the confused din of massive machinery – the burning headlands – the coke hearths, now if the night be stormy, bursting into sheets of flame, now wrapt in vast and impenetrable clouds of smoke – the wild figures of the workmen, the actors in this apparently infernal scene – all combined to impress the mind of the spectator wonderfully.

The manufacture of iron was a grim and at the same time an exhilarating and powerful spectacle. Seen from close quarters by contemporary observers, or from a more general

Ironmaking

perspective of its prodigious growth in output, ironmaking was a heroic endeavour, sentiments that rubbed off on the early historians of the industry. If there was one thing in which the Victorians had absolute confidence it was progress. Early historians were closely associated with the industry, men like Harry Scrivenor, erstwhile manager of Blaenavon Ironworks and author of *History of the Iron Trade* (1841), and Dr John Percy (1817–1889), lecturer on metallurgy at the Metropolitan School of Science and the Royal School of Mines, London, and author of *Metallurgy* (1864). They saw technological development and the industry's increasing productivity as its main historical themes. Percy was at the forefront of the burgeoning science of metallurgy and could see his own place in the larger scheme of things.

Scrivenor, Percy, and before them James Weale, identified a modern era of ironmaking based on the use of mineral fuel and the adoption of Henry Cort's puddling and rolling process for manufacturing malleable iron. The earliest period, extending back into prehistory, was the direct process of smelting in a bloomery to produce malleable iron using charcoal fuel. This was superseded from the late fifteenth century by the adoption of an indirect process, of smelting iron ore in a blast furnace with charcoal fuel to yield molten iron that needed converting to malleable iron at a finery. In the eighteenth century mineral fuel was adopted, specifically coke in the blast furnace and coal in a variety of refining techniques, before puddling became the most widely adopted process by the early nineteenth century. Both Percy and Scrivenor were writing in a period before Bessemer steel became established in the 1870s, coinciding with the decline of the wrought-iron industry in the final quarter of the nineteenth century, and providing the final main period of manufacturing malleable iron. These main periods can be summarised as follows:

1 A forge in South Wales, by Julius Caesar Ibbetson, 1789. For artists of Ibbetson's generation, it was natural to want to portray an industry that had become a phenomenon in its own right. (© Cyfarthfa Castle Museum & Art Gallery, Merthyr Tydfil)

1. The direct process, whereby iron ore was smelted to produce malleable iron in a single hearth, known as a bloomery, using charcoal fuel.
2. The indirect process, where iron ore was smelted in a blast furnace, producing pig iron that was converted to malleable iron at a finery, again using charcoal fuel.
3. The substitution of coke for charcoal.
4. Puddling and rolling, a coal-based method of refining coke-smelted pig iron.
5. Steelmaking by the Bessemer and open-hearth techniques.

Technological change has always been a fundamental issue to iron-industry historians. In the twentieth century historians broadened their concerns to take account of social, economic and political contexts. Some factors remain undisputed. The war-blighted eighteenth century created unprecedented demands for iron of many kinds at the same time as imports were inhibited. The Seven Years War (1756–1763), American War of Independence (1776–1783) and French Revolutionary and Napoleonic wars (1793–1815) each saw a marked upturn in the number of ironworks, the output of individual works and the industry's overall profitability. This growth was followed by a further surge arising from new civilian uses for iron, amounting to a peak period followed by a decline, the general model that has structured thinking about the iron industry as a whole, and of regions and individual ironworks. For example, the conservative instincts of the industry have been cited as the reason why newer firms pioneered new technology and out-performed their older competitors. Such an argument would account for the rise of South Wales in the decades before and after 1800, and its decline from the 1870s, when newer firms elsewhere invested more heavily in steel-making technology and were better placed to exploit raw materials. Economic historians have argued that the adoption of new processes was a function of their profitability. A deterministic application of this argument has been challenged, on the basis that quality of iron was equally as important as cost — but both arguments have been cited as the reason steel superseded iron in the late nineteenth century.

This general survey of the iron industry seeks to qualify and challenge some of the prevailing notions in the historiography of the industry. It has been customary to talk of an industrial revolution in the iron industry in the late eighteenth century. However, this rather loose phrase needs to be examined more critically: evidence of the forge trade shows that technological developments were established slowly and were used in conjunction with earlier technology. Study of workplace and business culture shows that new technology made little impact on the way the trade conducted its business or organised its labour force. By the mid-nineteenth century ironmasters continued to invest technical authority on a core group of skilled men, just as they had done a century earlier. Despite the potential of a national railway network, the industry continued to retain its strong regional character in the nineteenth century, a factor that has contributed to the strong tradition of regional histories of the iron industry. The iron industry can therefore be explained as a long historical process and not subject to abrupt forward leaps.

Historians and archaeologists have built their understanding of the industry on a linear model of technological development, largely drawn from the award of patents in the eighteenth and nineteenth centuries. The technological approach to iron industry history has, in effect, been to identify successful technology and then to interpret the preceding years as following a path to the ultimate destination. There is no evidence that technological change moves in so straightforward a way. It embodies an underlying notion of the industrial revolution as a triumphant forward march. In practice it has meant that the persistence of older technology has too often been deemed 'old fashioned', when a better explanation is needed. In particular, the model of linear progress has dogged our understanding of the development of forge technology in the second half of the eighteenth century. After 1750 several coal-based methods of refining pig iron were attempted, with varying degrees of success, before puddling was employed successfully in the 1790s. Twentieth-century knowledge of the chemical processes involved in refining pig iron has been transposed to eighteenth-century techniques, in the hope of explaining them and assessing their importance. The problem with this approach is its assumption that technology follows a linear course, and that earlier techniques were a more primitive version of later techniques. At its worst it introduces a patronising tone to the debate. But more importantly, it presents a misleading picture of technology as something performed by workmen under the technical supervision of what by the twentieth century might have been classed as the engineer. Official fees and stamp duties required to obtain a patent in the eighteenth century amounted to £70, excluding other gratuities payable, putting the system well beyond the reach of workmen. The ironmasters whose names appear on the patents brokered the new technology and hoped to profit from it, but they were not necessarily the originators of new processes.

One of the chief aims of this study is to rehabilitate the workmen, whose importance has been underestimated by technological historians, if not by social historians. Previous studies of technological change have stressed the importance of individuals – Darby, Cort, Neilson, Bessemer – but in doing so a whole layer of expertise behind them has been written out of the script. The trend began with nineteenth-century historians such as Samuel Smiles and John Percy who, given that they were writing when metallurgy was emerging as a scientific discipline and could see the impact of inventors like Henry Bessemer and the Siemens brothers, tended to construct the past in the image of the present. Since then, the emphasis on individuals has become part of a well-defined tradition of the heroic inventor. Before Bessemer, however, ironworking was a more empirical and collaborative activity that is worth re-emphasising.

It is also the case that the experiences of a handful of dominant figures did not reflect the experiences of most ironmasters or iron workmen. Ironmaking was not simply an economic or technical activity, it was a sub-culture in its own right, with its own traditions, prejudices and hierarchies. A general history of ironmaking therefore needs to re-establish the wholeness of the industry as the essential unit of analysis. In order to achieve these aims, it must consider the economic and technological developments within a broad cultural framework that encompasses the world of ironmaster, workman and merchant. The world of the merchant is particularly significant because it has generally been missing from previous studies of the iron industry. Neither economic, technological or cultural approaches have given adequate consideration to the influence of the market in technological change. That the market was a passive receiver of new technology has often been taken for granted.

What we need is a re-examination of technological change in the iron industry and the methodology of studying it. Economic determinism and the certainties of linear progress structure the history of ironmaking around individual technological breakthroughs, and imply that change was precipitated by events rather than processes. Gradualists have overtaken revolutionaries in most branches of industrial history but to date the iron industry has remained largely impervious to such reassessment. What is needed is a critical analysis of the assumptions that lie behind the history of the industry.

2

THE BLOOMERY

The discovery and dissemination of ironmaking technology from the Near East is a classic example of diffusion. The earliest known iron artefact is a dagger discovered at Alaça Hüyük in central Turkey, dated 2500–2300BC. It was found in a grave, in a context with high-prestige bronze objects. Ironworking had reached Greece before 1200BC. Like the Turkish prototypes, early finds from Greece are small items from graves – brooches, rings, knives – although more ambitious swords were forged by about 1050BC. The technology reached Germany and France before 700BC, and Britain half a century later, probably coming to Britain by the Atlantic seaways, on the basis that excavated furnaces in western Britain appear to be the earliest.

Although iron is the most common of the metallic ores, it is not as easy to smelt as copper, which is why copper and bronze were used before iron. By contrast, the widespread availability of iron ore explains why iron overtook non-ferrous metals in importance once the technology had become established. Iron may first have been produced by accident, when iron ore found its way into the charge of a copper-smelting furnace. Metallic iron was probably first noticed as a by-product of copper smelting whose value eventually came to be recognised in its own right. The close archaeological association of bronze and early iron objects suggests that the early iron smiths were also proficient in bronze manufacture, and that 'smith' denoted someone who worked in metals but without making a further distinction. The widespread deposits of iron ore ensured that the production of iron had a wide distribution, but there was nevertheless specialisation. In later prehistory the region with the most advanced iron industry, forging high-quality iron goods for a wealthy market in Italy, was the Alps. It is an early example of how technology thrives in favourable market conditions.

The technology of iron smelting was relatively simple – the difficult part was the technique. Until the late medieval period in Britain iron was smelted in a single direct process at a bloomery. A small bowl or shaft furnace was essentially an enclosed combustion chamber that allowed waste gases to escape from the top and iron and slag to be drawn off from the bottom. Fuelled with charcoal and fanned by hand-operated bellows, the furnace reached temperatures of up to 1,200C which, although it is below the melting point of iron, was hot enough for the slag to melt. Combustion of charcoal gave off carbon monoxide which gradually combined with the oxygen in the ore to leave iron in a metallic form, that coalesced into spongy lumps inside the furnace. At the same time, silica in the iron ore mixed with iron oxide and melted, leaving an iron-rich slag known as fayalite (iron silicate). Slag could either be tapped from the furnace (known as tap-slag) or, in bowl-shaped furnaces, remained within the furnace and is known as 'furnace bottom'. On removing the mass of iron, known as the bloom, at

Ironmaking

2 A late-medieval bloomery shown in Agricola's *De Re Metallica* (1556). It shows raw material being broken down to the correct size for the hearth at the top, from which iron was taken to the hammer at the bottom.

white heat from the furnace, it was hammered with repeated blows in order to shape the iron into a manageable block and to expel the liquid slag trapped in the interstices of the metal. In the Roman period, the earliest period for which there is good evidence of production, a bloom weighed 7–9kg, rising to 14kg by the fourteenth century AD. In practice, the bloom needed to be returned to the furnace and hammered several times before all the extraneous material had been removed and iron of tolerable quality was produced to be worked elsewhere into tools, weapons or other implements. Smiths' hearths were similar to bloomeries and in practice it can be difficult to interpret which process was carried out at a particular site. The presence of slag is usually the decisive factor. In practice, however, much of the evidence of the early iron industry is found as slag or iron ore without the presence of a furnace.

Iron in prehistoric and Roman Britain

Iron ore has been found in most districts of Britain and until the late medieval period it was smelted in areas not usually associated with the iron industry, like Essex, Cornwall and Skye. Until the medieval period the ore was almost invariably dug from the surface and smelted nearby. Iron ore is found in various forms. Carbonate iron ores are the most common in Britain, and were found either as nodules in the Weald and in the coalfields, or as a sedimentary deposit (also known as spathic iron ore). Roasting the ore had the effect of driving off the carbon in the form of carbon dioxide, leaving ferrous oxide suitable for smelting. Hematite is ferric oxide, much sought after during the industrial period as it contained comparatively low traces of phosphorus. It is chiefly found in Cumbria, and especially the Furness district. Limonites are hydrated iron oxides, the largest deposits of which are in the Forest of Dean and the southern end of the South Wales coalfield. Bog iron ore is found across northern and western Britain. It forms in wet conditions where iron-bearing waters meet organic material. In Sweden it has been dredged from the bed of lakes, while in Britain it occurs beneath moorland turf.

Direct evidence of smelting is usually confined to the furnace bases, because furnace superstructures were temporary and have not survived. One of the best-preserved prehistoric bowl furnaces was found in Rudh' an Dunain cave in Skye. It was built of stone slabs, probably with clay to plug the gaps, with internal dimensions of 0.38m by 0.23m. It is dated to the fourth century BC. Iron slag was found within the furnace, and associated levels yielded birch and willow charcoal. Most furnaces, however, were less solidly built. A similar construction belonging to the Roman Iron Age was found in Constantine's Cave in East Fife. At Brooklands, near Weybridge in Surrey, iron was smelted over a period of five centuries and evidence of twenty-one furnaces used over that period has been found. The most complete had a bowl-shaped base dug into the ground, a superstructure of red clay, and a channel for tapping the slag. Most of the furnaces had internal diameters of only 0.3–0.35m, although six larger furnaces had diameters of 0.6m and may have been used for secondary forging. The most common form of slag found on the site was 'furnace bottoms'. Iron Age settlements at Hengistbury Head in Hampshire, Glastonbury Lake Village in Somerset and St Eval and Chysauster in Cornwall have all yielded iron slag, suggesting the presence of smelting.

Iron artefacts are rare before the Roman period. Iron was at first an exclusive rather than a commonplace material. Its adoption was hindered by the low level of economic development, at least until the first century BC. The rapid expansion of the industry in the century preceding the Roman invasion was concentrated in the Weald, probably under the influence of Belgic invaders, for whom iron was an integral part of the central European economy. Julius Caesar, who visited Britain in 55 and 54 BC, and the Greek geographer Strabo, writing half a century later, both mention iron as one of the industries of Britain.

The most prolific source of prehistoric iron artefacts is hoards. At Llyn Fawr, just north of the Rhondda valley in upland Glamorgan, a hoard from the seventh century BC included bronze cauldron, axes and sickles, an iron sword and two iron sickles. The sickles and sword are iron versions of known bronze objects and suggest strongly that the earliest blacksmiths were also bronze smiths. They were probably full-time specialists enjoying elite patronage. Another substantial early Iron Age hoard was found at Llyn Cerrig Bach, Anglesey. Items appear to have been thrown from a rock platform into a lake, and included military equipment like spears, swords and shields, as well as harness ware and chariot fittings. Julius Caesar wrote of the Celtic practice of gathering the spoils of battle to make dedicatory offerings. There had also been a long tradition of depositing valuable artefacts – and people – in watery locations. Conspicuous disposal of valuable items like swords suggests that metal objects had special properties associated with individuals, and that the discarding of materials was a symbolic act, perhaps following the death of their owner. It also demonstrates that iron was valued far beyond its immediate utilitarian function. Most discarded iron objects have been of military character. Iron swords have been found at a number of sites. Tapering sword-shaped bars have also been found that have been interpreted as the currency bars mentioned by Julius Caesar as a form of exchange in Britain in addition to gold and bronze coins. Such bars may have been the primitive initial bars from which swords were later forged. Other uses of iron were for chains, axes and knives. Rarely has evidence been found that iron was a decorative metal, although iron torcs have been found at Spettisbury, Dorset.

Iron became an important industry in Roman Britain. Roman occupation brought a much larger economy, more efficient administration and improved technology. At Ashwicken, Norfolk, is a second-century iron-smelting site where an almost complete furnace was found, with an internal diameter of 0.3m and a height of 1.4m. A single opening at the base was used to remove the bloom, tap the slag into a channel and a shallow pit, and to blow the furnace with bellows. Apart from the successful use of larger furnaces, innovations of the Roman period included roasting hearths. Iron ore was burnt in order to drive off moisture, reduce the damaging sulphur content of the ore and to break it up into manageable sizes for the furnace. Ore was still easily available from surface workings, although an adit of Roman date has been discovered at Lydney in the Forest of Dean.

Several smithing furnaces for working up iron into artefacts have been found at Corbridge, Northumberland, which suggest that the military purchased iron from independent smelters, although evidence of smelting has also been found at Galava fort near Ambleside. In the more developed and pacified south Britain, the industry saw considerable expansion. The peak of the Roman iron industry was from the later first

3 Ore roasting, from Agricola's *De Re Metallica* (1556).

century AD to the middle of the third century, when there is evidence of a decline in smelting in the Weald and abandonment of some of the key sites. The rapid growth of the Wealden industry probably represented more than just an expanded market to be exploited by native entrepreneurs, but was a centrally planned and administered operation. When Britain was absorbed into the Empire, its mineral reserves became the property of the state. It has been argued that the iron industry in the eastern Weald was under the control of the *classis Britannica* – the British fleet during the Roman occupation and under the command of the Roman army – and by implication that British iron was exported for use in the north-western provinces of the Empire. Elsewhere sites were probably leased to entrepreneurs and the work was undertaken by self-contained communities.

Bardown and Beauport Park are two Wealden ironmaking sites that were probably under military control. At Bardown, near Wadhurst, smelting began in the second century AD. Refuse in the form of slag and furnace debris, and pits from which ore was dug, are close to a related settlement. Archaeologically, the site is typical in that its remains constitute a slag heap, a working area higher up the slope, and an ore source. It has been calculated that seven or eight furnaces had an aggregate annual output of 40–45 tons, while the distinct layering of material in the waste dumps suggests an annual cycle of ore digging, timber felling, charcoal burning, smelting and forging. Among the buildings excavated was a timber-framed military style barrack block that would have housed up to forty men. In the early third century smelting declined but began at a series of satellite sites, including Holbeanwood a mile north of the settlement. The most likely reason for this shift was the availability of ore. The same phenomenon has been interpreted at Ashwicken.

Apart from tools and implements such as knives, chisels, scythes, swords and javelins, many other items required iron components, such as hinges, locks, nails, horseshoes and barrel hoops. A hoard of over 5 tons of nails was found at the first-century Inchtuthil fort in Scotland. Larger items which required several blooms to be welded together included anchors and iron beams for bath houses. The latter were placed above a furnace chamber to support a bronze water heater. The largest of such beams found in Britain, at Catterick, was over 1.7m long. Roman technology also included superior techniques of heat-treating wrought iron to produce steel. Steel has a small proportion of carbon, less than 1 per cent, which became infused into the surface of the metal when iron bars were repeatedly heated in charcoal hearths. Both the short stabbing sword, the *gladius*, and the long sword, the *spatha*, were forged from carburized iron. The Romans also introduced pattern welding to Britain, whereby a sword was forged from several thin steel strips twisted and welded together. The welding of several bars which had a hard steel surface allowed a stronger finished product composed of layers of iron and mild steel. The pattern, although it was devised to achieve the maximum distribution of steel within the structure of the sword, became a decorative feature, a status symbol, and proof of the quality of the weapon.

Iron in medieval Britain

The iron industry declined after the collapse of the Roman administration in the fifth century, largely because the economy it served contracted. Revival was slow and there was no significant technological development until the late middle ages. A smelting

Ironmaking

site at Ramsbury, Wiltshire, dated to the late eighth century, had furnaces little different and no larger than Roman furnaces. By the eleventh and early twelfth centuries some furnaces were slightly larger. Excavated furnaces at West Runton, Norfolk, and Lyveden, Northamptonshire, were 0.5m in diameter.

Until the second millennium AD there are few documentary sources relevant to the iron industry. Manorial and lay estate records document the mining, smelting and sales of iron, although in only a few cases have they been linked to extant archaeological sites. Where they occur, documentary sources tell us much that would otherwise be difficult to discover from archaeology alone. One of the key sources for medieval ironworking is found in the accounts recording the operation of a bloomery at Tudeley, near Tonbridge in Kent, for the period 1329–54. The estimated annual output of the bloomery varied between 1.5 and 3 tons, each smelt producing a bloom of up to 14kg. The ore was roasted before smelting. Half the running costs of Tudeley furnace were in the acquisition of charcoal, despite the fact that the wood came from the owner's own estate. The chief factor limiting output was not fuel but the source of power. At Tudeley the greatest wage expenditure was not on smiths or furnacemen, but for the team of four blowers, led by the master blower John Tubbe, who manually operated the bellows.

Blooms were formed into standard-sized bars for sale, where they were worked up into finished products at smithies. At Roman smithies the hearth was inside a substantial building. The same was true in the medieval period, for example at Waltham Abbey, where a stone-walled building of the twelfth century has been excavated. It had two hearths fanned by bellows, would have needed a water bosh for cooling the iron and the tools, and also had a small lead smelter. The market for iron products grew in response to economic

4 A locksmith's workshop from a medieval German *Book of Trades*, showing the bellows-blown hearth, anvil and work bench.

The bloomery

growth. As trade expanded, there was a greater demand for horseshoes, and for nails and other fittings used in ships and wagons. Improved techniques increased the range of iron products. For example, the ability to work the iron cold in a similar manner to hard wood, by sawing, filing and drilling, and known as the locksmith style, was used for locks and architectural ironwork. Nevertheless, Britain became a net importer of iron, which it had not been in Roman times. Imported iron was shipped from France, Spain, Sweden and Germany, often of higher quality than the home-produced alternative. The highest quality appears to have been Swedish iron known as 'Osmunds'. Iron was hardened to produce steel for cutting edges in knives, axes, chisels and spades, most of which was imported from Sweden, Russia and Spain. The Royal building accounts for 1278−79 show that English iron was used for manufacture of nails, horseshoes, wedges, spades and pickaxes, but iron for siege engines and weapons was imported from France, Spain and Sweden. For those who could afford it, the best armour was made from Spanish iron.

Smiths appear to have been men of high status and a few of their names are known. In 1294 Thomas of Leighton (probably Leighton Buzzard) was paid £12 for erecting an iron screen at Westminster Abbey, London. Contemporary ironwork of a similar style, and therefore probably also by Thomas, survives in churches at Turvey and Leighton Buzzard in Bedfordshire. The ironwork of *c.*1240 on a door at St George's chapel, Windsor Castle, is stamped 'Gilbertus'. The tomb railings that became increasingly popular in the late medieval period were a sign of status and are among the earliest freestanding structures of iron.

Waterwheel technology came to the iron industry in the second millennium. The technology was known in Europe during the Roman period and its use gradually increased thereafter. By the beginning of the second millennium there were over 5,000 corn mills in England alone. The use of water power to work the hammer or the bellows in the iron industry probably originated on the continent. In the twelfth and thirteenth centuries water-powered hammers were employed at iron mills in Spain, France and Germany. The earliest documented use of a waterwheel for the iron industry in Britain was at Chingley in Kent, in the early fourteenth century. The site has been excavated and proved most likely to have been a forge for hammering blooms into bars, not where ore was smelted. Foundations for a large hammer were found, and the hearths here may also

5 Iron railings and gates at Farleigh Hungerford Castle chapel, Somerset. The railings surround the tomb of Sir Thomas Hungerford and were made between 1465 and 1470, at a time when the use of decorative ironwork was a sign of high status. (© author)

have been blown by water-powered bellows. It has been calculated that 8hp was required to work a hammer and 1hp to blow the bellows for a furnace, but the importance of water-powered bellows only increased after the Black Death, when the decrease in population put a premium on labour costs.

The introduction of water power had two main consequences. In future, bloomeries and forges could only be set up where sufficient water was available, and therefore had to complete with other industries such as corn mills and fulling mills. Secondly, improved output was possible because the furnaces could be built larger. An especially well-documented bloomery of the later medieval period is Byrkeknott in County Durham, built in 1408 at Weardale on land owned by the Bishop of Durham. The site is probably what later became a corn mill known as Harthope Mill, which was excavated in the 1950s. Whereas only a single hearth had been in use at Tudeley, at Byrkeknott iron was smelted in the bloomery and reheated to allow prolonged hammering in a second separate hearth, known as a string-hearth. String-hearths have been identified on archaeological sites by their absence of provision for slag tapping. A waterwheel of about 5m diameter drove a hammer and also bellows for both bloom-hearth and string-hearth, although the bellows were probably operated manually when the water supply was insufficient. In its first year of operation, 24.3 tons of iron were produced at Byrkeknott and individual blooms weighed up to 87kg.

Increased output of the water-powered bloomeries required larger-scale extraction techniques. Deep mining began in the medieval period, although evidence for it has largely been destroyed by later workings and mine workings are difficult to date on archaeological evidence alone. Compared to the later blast furnaces, water-powered bloomeries remained small, although few have been excavated. The best examples are in the north of England where bloomery technology continued in use until the early eighteenth century to serve the comparatively small market for iron in the region. At Rockley Smithies in South Yorkshire, a bloomery that had worked in the late sixteenth century was rebuilt around 1600 and continued to operate until around 1640. It had a bloom-hearth and two string-hearths but they remained small examples of the bowl-type furnace, with an internal diameter of only 0.8m. Fragments of two 3.3m diameter waterwheels were found. The bloomery had separate waterwheels to power the bellows for the bloomery and string-hearths but not, apparently, for the hammer. Muncaster Head, Cumbria, was built in 1636 and was one of the last bloomeries to be established in Britain. It continued working until 1720. The site was identified and excavated in the 1960s. Evidence of water power, in this case for the hammer, confirmed what was known from documentary sources. Slag and furnace bottoms confirmed the presence of a bloomery, but no structural evidence was found for a string-hearth or hammer.

As the capacity of furnaces increased, so the industry became confined to those regions with the best reserves of ore. The proliferation of iron objects also saw the emergence of regional specialisation. Country blacksmiths remained important for routine work such as shoeing horses and maintenance of agricultural tools. The Midlands had become an important ironworking district during the medieval period and was a centre for the manufacture of bridle bits, horseshoes and nails by the early sixteenth century. Belbroughton, Chaddesley Corbett and Clent in Worcestershire specialised in scythe making, Wolverhampton in lock making, Walsall in harness ware. Blade manufacture, which used water-powered grinding wheels, had become a Sheffield specialisation by the end of the medieval period. The regional structure of the iron industry was already forming by the time that blast furnace technology came to Britain.

3

THE BLAST FURNACE AND FINERY FORGE

From the late fifteenth century the direct reduction of ore in a bloomery was superseded by the indirect process – smelting in a blast furnace, then refining at a forge. The overlap between the two processes in Britain was a long one, as it had been earlier in continental Europe. The blast furnace was less an invention than the product of a long period of development. Of the two main structural forms of the bloomery, tall shaft bloomeries were not favoured in Britain but had a more widespread use elsewhere in Europe. In a shaft furnace, iron remained in contact with the charcoal at high temperature long enough for the iron to attract carbon. In this form it had a lower melting point than pure iron, and so in the shaft bloomery, molten iron may first have been produced by mistake. A use was found for the molten iron by pouring it into moulds – cast iron was an established part of the ironworkers' repertoire in Europe by the end of the fifteenth century. A more fundamental development was to refine cast iron to form wrought iron, which was also a development of bloomery skills. Cast iron has a carbon content of 3–4 per cent, making it too brittle to be malleable under a hammer. It consequently required further refining to forge bar iron that was sold to smiths for working up into finished products. This secondary process was undertaken at a 'finery' forge, evidence of which has been discovered in Sweden for as early as the twelfth century. By the middle of the fifteenth century it had become well established – a description of a blast furnace and finery in Tuscany was made by Antonio Averlino Filarete in around 1460. An additional process, of heating refined iron ready for hammering into shape ready for sale, was undertaken at a chafery, counterpart of the string-hearth in the bloomery phase of ironmaking.

Both the blast furnace and the forge relied upon water power, the availability of which influenced the relationship between the two. As outlined below, some sixteenth- and seventeenth-century forges were built in close proximity to blast furnaces. Such an arrangement was only possible where there was ample water supply, or where the furnace and forge operated in relay. However, in cases where iron was allowed to cool at the end of one process and needed to be re-heated for the next process, transport of semi-finished iron was possible and in practical terms often desirable. Geographical separation of furnace and forge was a consequence of this as was, on a smaller scale, separation of finery and chafery. Bloomeries had also relied upon water power and therefore provided opportunities for conversion to blast furnaces or forges. Excavated sites where the finery superseded the bloomery include Woolbridge, Mayfield and Brookland in the Weald. Other sites have been identified by documentary and archaeological evidence in Furness, where finery forges superseded seventeenth-century bloomforges at Backbarrow, Cunsey, Spark Bridge, Coniston, Force, Hacket and Burblethwaite. Cunsey bloomforge is first recorded in 1623 but, after the construction of a blast furnace further downstream in 1715, it became a finery forge until the furnace ceased work in 1750. Backbarrow bloomforge was built in 1695, but was converted to a finery and chafery forge some time after the Backbarrow blast furnace was built in 1712, and continued working to the mid-nineteenth century.

The geography of iron smelting

By the early eighteenth century the British iron industry had formed into five principal regions: the Kent and Sussex Weald; the Forest of Dean and Monmouthshire; the Midland counties of Staffordshire, Shropshire, Worcestershire and Warwickshire; and Yorkshire and Derbyshire. Smaller but not insignificant regions included the Furness district of Cumbria. Each of these regions had reserves of iron ore and suitable woodland for manufacture of charcoal.

There is no evidence that blast-furnace technology in Britain was the result of an independent evolution. A bloomery at Timberholme, on the estate owned by Rievaulx Abbey, probably produced cast iron but there is no evidence for sustained production. By contrast, there is good evidence that the technology and expertise for significant production of molten iron was imported. The first blast furnace erected in Britain was in 1496 at Newbridge in the Ashdown Forest district of the Weald. It stood on the site of a former bloomery. Accounts and an inventory of 1509 show that the furnace worked in conjunction with a finery and chafery. By 1520 there were still only two blast furnaces in the Weald, as well as numerous traditional bloomeries. Subsequently there was a quickening in the pace of development. By 1548 there were fifty-three furnaces and 'iron mills' in Sussex. They employed a high proportion of immigrant workmen, mainly from the *pays de* Bray in Normandy, who brought and disseminated furnace, foundry and forge skills in Britain. By 1574 the fifty-two blast furnaces and fifty-eight finery forges outnumbered bloomeries in the Weald, the industry now having expanded into Kent and Surrey. Industrial growth went hand-in-hand with woodland management. Wood was a renewable and valuable economic resource where it was coppiced, a practice well developed in the Weald in the sixteenth century. Charcoal was made from coppice shoots or from the 'lop and top', i.e. the branches of larger felled trees, although there were competing claims on wood, notably for domestic fuel. It has been calculated that in the sixteenth century 4,000 acres of well-managed woodland would be sufficient to support a blast furnace and finery forge in the Weald.

From the second half of the seventeenth century the Wealden iron industry concentrated on cast iron wares, and maintained its position as the nation's chief producer of ordnance until the 1770s. The seventeenth century was nevertheless the beginning of a protracted decline. The region's iron industry was said to have shrunk to thirty-six furnaces and forty-five forges in 1653, and by the second decade of the eighteenth century there were only thirteen forges in Kent, Surrey and Sussex. At the same time there were fourteen blast furnaces, although none smelted more than 350 tons of pig iron per year at a time when modern furnaces in the Forest of Dean and elsewhere could produce over 600 tons.

The blast furnace and finery forge

The Forest of Dean possessed some of the best-quality iron ores in Britain, but the blast furnace did not supersede the bloomery here until the seventeenth century. Three blast furnaces had been built near Dean's woods in the late sixteenth century, the earliest of which was on the Welsh side of the River Wye at Whitchurch. Subsequently four blast furnaces were erected in the Forest after 1612 by William Herbert (1580–1630), the Earl of Pembroke, that became known as the King's Ironworks. Three of them – Lydbrook, Parkend and Soudley – supplied a finery forge built downstream from the furnace. The fortunes of the Forest of Dean iron industry in the seventeenth century are comparatively well documented because of disputes with the Crown over the cutting of wood for charcoal, and the hiatus surrounding ownership in the middle of the century. In 1634 there were eleven blast furnaces and eleven forges in and immediately around the forest. Control of Dean's iron industry soon came into the hands of the Catholic Sir John Winter when he purchased the Crown forests, but they were confiscated when the king's authority collapsed in 1642. Erratic administration of the forest in the Civil War was followed by a period of further development under Major John Wade, chief administrator of Dean during the Commonwealth, before reverting to the Crown on the Restoration. The industry was subsequently impeded by accusations of badly managed woodland resources, and in 1673 furnaces at Parkend and Lydney were demolished after access to local wood was denied. The number of Dean ironworks subsequently declined, partly because the Crown decided to sell cordwood for charcoal to ironworks outside of the Forest. Chief of these were the ironworks owned by Paul Foley, whose wood buyers, John Wheeler and Richard Avenant, made a contract in 1692 for wood that raised £20,000 for the Crown in seven years. By the second decade of the eighteenth century there were six blast furnaces and thirteen finery forges in Gloucestershire.

Blast furnaces were introduced into the ironworking district of South Yorkshire and North Derbyshire in 1573–74. During the seventeenth century ownership of furnaces and forges became concentrated into the hands of a group of gentry ironmasters, notably the Spencer and Cotton families. By the second decade of the eighteenth century there were four blast furnaces in Derbyshire, six in Yorkshire, with four and nine finery forges respectively. In the iron industry centred upon Furness in south and west Cumbria the bloomery remained in use throughout the seventeenth century. But the region was not isolated from technical developments. The quality of its iron ore was well known and it was shipped to a number of blast furnaces in the later seventeenth century, including Lawton and Vale Royal furnaces in Cheshire, and even to the Forest of Dean. The earliest blast furnace in the district, at Cleator, was built in 1694.

In the Midlands, blast furnaces were built in around 1563 at Cleobury Mortimer, Shropshire, by Robert Dudley (1532–1588), later Earl of Leicester. In 1564 a former bloomery at Lizard, near the Shropshire/Staffordshire border, was converted to a finery forge by George Talbot (1528–1590), Earl of Shrewsbury, and was supplied with pig iron from a furnace built by Talbot at nearby Shifnal in Shropshire. Although the district of south Staffordshire and north Worcestershire, later to be known as the Black Country, was well-endowed with reserves of iron ore, development was impeded by poor water supply. It had comparatively few furnaces in this period, only four by the second decade of the eighteenth century.

Ironmaking

Furnace and forge in operation

The technology of blast furnace and forge differed little across Britain. In a blast furnace, iron ore was charged into a tall stack inter-layered with charcoal and limestone and yielded molten iron. The limestone, containing calcium carbonate, was useful because it combined with the stony constituents of iron ore, mainly sand, silt and clay, to form a fluid slag that solidified as soon as it was tapped from the furnace. To attain a consistently high temperature in the furnace – over 1,200C – a blast of air was provided by water-powered bellows. The iron was run into moulds prepared on the casting floor, usually in the form of a line of ingots known as pigs.

Blast furnaces became progressively larger than the bloomeries they superseded, although their scale was limited by the comparatively fragile nature of charcoal – if the layers of ore and charcoal in the furnace were too heavy the charcoal would crumble and be rendered less effective. The blast furnace stack was encased within an outer masonry casing, and an inner face of heat-resistant stone. Georgius Agricola depicted a blast furnace no bigger than 2m high in his mid-sixteenth-century treatise on metallurgy, a direct descendant of a shaft bloomery. Excavated Wealden blast furnaces of the sixteenth century were clearly much larger as they were between 5.5m and 8m square at the base. Lydbrook furnace in the Forest of Dean was described in 1635 as 23ft (7m) square at the base, with tapering sides 23ft (7m) high. Because furnaces were charged at the top and tapped at the base, they were usually built against natural banks.

6 An early blast furnace, clearly derived from a shaft bloomery, illustrated in Agricola's *De Re Metallica* (1556).

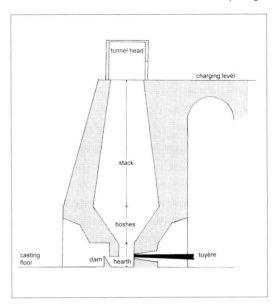

7 Cross-section through a blast furnace, showing its main components.

The iron ore was usually calcined – burned to remove some if its impurities – close to the source of the ore, in order to prevent transport of unnecessary waste material. It was then beaten into small pieces, as was the limestone. The raw materials were gathered at the site and were stored in sheds to keep them dry. At the top of the furnace was a 'bridge house' where the iron ore, charcoal and limestone was tipped in from baskets. The furnace required about an equal number of baskets of charcoal and ore. At the base was a 'casting house' where the iron was cast into pigs or moulds. Air for the blast was provided by a pair of bellows, which at Lydbrook was powered by a waterwheel 23ft (7m) in diameter. A contemporary description of the blast furnace built at Leighton in Furness in 1713 specifies that the bellows were each 7½ yards (6.9m) long and 1½ yards (1.4m) wide. Power was provided by a 30ft (9.1m) diameter waterwheel, turning an axletree 36ft (11m) long, on which cams alternately depressed the bellows. The bellows were raised again by heavy counterweights at the rear, and the system worked in such a way that the furnace received a constant blast of air directed through a pipe known as a tuyere. At Leighton the furnace was tapped every twelve hours, yielding between 16 and 22cwt (0.8–1.1 tons) of pig iron at each tapping. Half a century earlier John Ray reckoned that the output of a Wealden blast furnace averaged about 8 tons a week.

Once a blast furnace had been lit it was kept constantly in blast. John Ray described blast furnaces in the Weald in 1674 as being lined with sandstone at the base, and brick above. It was reckoned that a sandstone hearth would last no more than forty weeks before it was so worn away that it needed replacing, and that when new a hearth had a capacity for 700lb (318kg) of molten iron, but by the end of its working life it had a capacity of 2,000lb (908kg). Limited availability of water restricted blast furnace operations to 'campaigns', usually of between six and nine months. In practice there was much more variation. Between 1621 and 1625, for example, only two campaigns were undertaken at Cannop furnace in Gloucestershire, of twenty and nineteen weeks. From 1741 a campaign at Gloucester Furnace in Sussex lasted just over three years. Between campaigns it was necessary to manage the furnace to prevent it getting damp. John

Ironmaking

8 An eighteenth-century blast furnace, showing the charging from the bridge house above, and the waterwheel powering the bellows. On the right side of the building are the protruding counterweights that raised the bellows depressed by the action of the waterwheel.

Hanbury (1664–1734), the Pontypool ironmaster, had a thatched roof built over the top of his furnace and in 1704 was considering building a ventilated brick cupola to allow a fire to remain lit in the base of the furnace to keep it warm.

One of the advantages of the blast furnace was that the molten iron could be cast into moulds to form finished products such as firebacks or cannon shot. From the outset Leighton blast furnace produced cast-iron wares – chimney backs, garden rollers, pots and pans – by running molten iron into a large cauldron from which it was ladled into moulds on the casting house floor. The importance of the arms trade was established in the sixteenth century when it became possible to cast cannon with iron rather than bronze, which was more expensive. William Levett of Stumbletts and Buxted furnaces in Sussex was described as the king's gunstone maker in 1541. Gun shot was already cast in the Weald and the two components formed the bulk of the British ordnance trade. Cannon were cast direct from the blast furnace. Excavation of Wealden blast furnaces has yielded evidence that in the sixteenth and seventeenth centuries large items such as cannon were cast in moulds set vertically in pits dug in the casting house floor. At Pippingford and Scarlets were well-preserved waterproofed, timber-lined casting pits, of which Pippingford, dated around 1700, was 4.5m deep and about 1.5m in diameter. The mould was set on a table in the pit, and one of the greatest difficulties was in lifting the casting out of the pit, since no evidence of cranes or hoists was found. At Pippingford an unfinished cast-iron cannon was discovered, over 1.6m long. The find is remarkable because iron had an obvious scrap value, either for charging into the blast furnace, or by treating it as any other pig iron and converting it to wrought iron. Hollow-cast guns needed to be finished to the correct internal bore. This was done at on-site boring mills, of which Pippingford is among the best archaeological survivals. The cannon was mounted on a wooden trolley, which was slowly advanced against revolving cutters, powered either by a waterwheel or treadmill.

The blast furnace and finery forge

At the forge, pig iron was melted in a charcoal-fuelled finery hearth and was kept in motion by stirring it with an iron bar. This ensured that the iron was uniformly exposed to a blast of air, provided by water-powered bellows, that allowed oxygen to combine with carbon in the pig iron to leave the iron in a pure state. The process depended upon the skill of the workman in ensuring even exposure to the blast of air, and judgement on when the iron was ready. The hearth was open-fronted, and had a tall stack above it. In order to provide enough iron to keep a hammer and a chafery working, two finery hearths was the most common arrangement.

Once the metallic iron had formed it was removed and taken to a water-powered tilt hammer. Repeated blows removed much of the slag trapped in the interstices of the metal, a process known as shingling, and shaped the iron into a bloom. The iron required several heats in the finery before it was judged sufficiently refined to be sent to the chafery. The chafery hearth was similar to the finery hearth but larger. In it the iron was brought to a white heat, sufficient for it to be drawn into a bar under a hammer, ready for sale.

9 An eighteenth-century finery. The furnace is fanned by a pair of bellows on its left side; pig iron is fed from the rear. When the iron is taken from the furnace, it is first beaten with a sledgehammer to remove slag and to shape it sufficiently for it to be placed under the forge hammer..

10 An eighteenth-century forge hammer, under which iron is shingled to form a bloom; the hammer is water-powered. On the right is the shaft, or axletree, turned by the wheel, which has cams attached at its end. The cams lift the hammer head, allowing regular hammer blows.

Ironmaking

The geography of finery forges

Whereas the geography of smelting was influenced by the location of raw materials, the geography of the forge was strongly influenced by the market for wrought iron. The biggest concentration was in the Midlands, serving the regional manufacturing industries. Ever since the introduction of blast furnace and finery technology to the Midlands, forges were far more numerous than furnaces in Staffordshire and Worcestershire. The densest concentration of forges in the area, and in Britain, was for 12 miles on the River Stour upstream of its confluence with the River Severn to Stourbridge. Nevertheless, Staffordshire and Worcestershire forges were unable to meet the demand for bar iron in

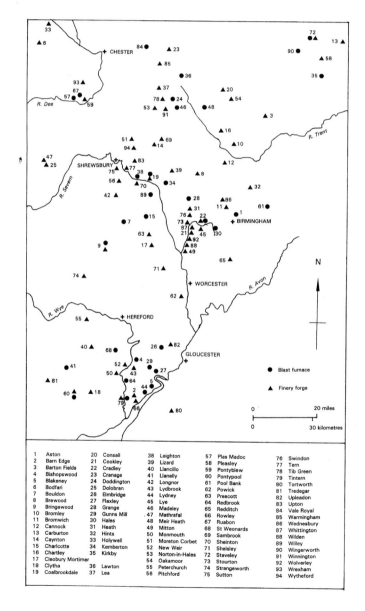

1	Aston	20	Consall	38	Leighton	57	Plas Madoc	76	Swindon
2	Barn Edge	21	Cookley	39	Lizard	58	Pleasley	77	Tern
3	Barton Fields	22	Cradley	40	Llancillo	59	Pontyblew	78	Tib Green
4	Bishopswood	23	Cranage	41	Llanelly	60	Pontypool	79	Tintern
5	Blakeney	24	Doddington	42	Longnor	61	Pool Bank	80	Tortworth
6	Bodfari	25	Dolobran	43	Lydbrook	62	Powick	81	Tredegar
7	Bouldon	26	Elmbridge	44	Lydney	63	Prescott	82	Upleadon
8	Brewood	27	Flaxley	45	Lye	64	Redbrook	83	Upton
9	Bringewood	28	Grange	46	Madeley	65	Redditch	84	Vale Royal
10	Bromley	29	Gunns Mill	47	Mathrafal	66	Rowley	85	Warmingham
11	Bromwich	30	Hales	48	Meir Heath	67	Ruabon	86	Wednesbury
12	Cannock	31	Heath	49	Mitton	68	St Weonards	87	Whittington
13	Carburton	32	Hints	50	Monmouth	69	Sambrook	88	Wilden
14	Caynton	33	Holywell	51	Moreton Corbet	70	Sheinton	89	Willey
15	Charlcotte	34	Kemberton	52	New Weir	71	Shelsley	90	Wingerworth
16	Chartley	35	Kirkby	53	Norton-in-Hales	72	Staveley	91	Winnington
17	Cleobury Mortimer			54	Oakamoor	73	Stourton	92	Wolverley
18	Clytha	36	Lawton	55	Peterchurch	74	Strangeworth	93	Wrexham
19	Coalbrookdale	37	Lea	56	Pitchford	75	Sutton	94	Wytheford

11 Ironworks in the Midlands, Forest of Dean and Welsh borders in the early eighteenth century.

The blast furnace and finery forge

the Black Country and Birmingham. In 1737 Abraham Spooner, a Birmingham iron merchant, claimed that 9,000 tons of wrought iron were consumed annually in the district. Imports accounted for 2,100 tons and, according to a national survey of 1736 the combined output of Staffordshire, Worcestershire and Warwickshire forges was 3,760 tons. Therefore over a third of the iron consumed in Birmingham came from outlying districts, including Shropshire, Cheshire, Mid and North Wales.

The underlying problem for the Black Country iron industry was insufficient sites with suitable water power. Where such sites existed, they had often supplied power to other industries. For example, five Stour valley forges – Greensforge, Heath, Swindon, Whittington and Stourton – are known to have been erected on the sites of former corn mills. This was not unique to the Midlands and a long list of similar cases could be cited. For example, two forges in Mid Wales, at Mathrafal (1651) on the River Vyrnwy and Dolobran (1719) were also formerly corn mill sites. Prime positions for the exploitation of water power appear to have been occupied by corn and other mills long before the expansion of the iron trade in the seventeenth century. It is arguable that the availability of such sites, usually when a lease was to be renewed, was an influential factor in the spread of forges in the Midlands and elsewhere. The reverse is also relevant, that when leases for forges or furnaces expired they were occasionally converted for use by other industries. Surviving examples are Gunns Mill in Gloucestershire, a seventeenth-century blast furnace that was converted to a sawmill in the eighteenth century, and Nibthwaite furnace in Cumbria, built in 1735–36 but converted to a bobbin mill in the 1840s long

12 A forge near Dolgellau, Gwynedd, by Paul Sandby, 1776. The small forge was built in around 1720 to serve nearby Dolgun blast furnace. It ceased work soon after 1800. (© Ironbridge Gorge Museum Trust)

after smelting had ceased. Access to charcoal seems not to have been a fundamental problem. In 1736–37 Abraham Spooner and Edward Knight argued for their own interests by claiming to Committees of the House of Commons that there was sufficient woodland in the Midlands to allow a significant increase in iron production.

Pig iron for Midland forges was purchased from further afield, especially the Forest of Dean. In 1715 4,950 tons of pig iron were produced in the Forest, but only 2,000 tons was refined locally. The surplus was sent to the Midlands, where 2,500 tons were refined in the Stourbridge region. The trade in pig iron was aided by the concentration of ownership in the iron industry, resulting in the dominant inter-regional concerns of the Foley followed by the Knight family.

The Foley family emerged as leading ironmasters in the Forest of Dean in the seventeenth century. The dynasty was founded by Richard Foley (1588–1657), succeeded by his sons Thomas Foley (1617–1677) and Robert Foley (1627–1677), and two sons of Thomas: Paul Foley (1650–1699) and Philip Foley (1653–1716). By 1672 Paul Foley held furnaces and forges in the Forest of Dean and shipped iron to Philip Foley's forges in the Stour valley. The brothers subsequently combined their interests and from 1692 numerous ironworks encompassing the Forest of Dean, Monmouthshire and the Midlands were managed jointly by two concerns: The Staffordshire Works and the Ironworks in Partnership. Co-partners in the Ironworks in Partnership included John Wheeler and Richard Avenant, both of whom had entered the iron trade as wood buyers in the 1650s, before taking on the management of Midland forges for Thomas Foley. It was a business empire that stretched from Meir Heath blast furnace in north Staffordshire to Tintern wire mills in the Wye valley. Shipments of iron between these works established an inter-regional trade in iron. The Stour valley was the focus of the partnership's forge trade, where slitting mills at Wilden, Cookley, Stourton and Wolverley, all working in conjunction with finery forges, supplied iron for nail manufacture.

The Ironworks in Partnership subsequently declined and after 1711 withdrew to the Forest of Dean. In the Midlands it was superseded by another dominant ironworking dynasty, the Knight family. Richard Knight (1659–1745) was probably born in Madeley, Shropshire, and is said to have begun his career at a Coalbrookdale forge. Subsequently he took over the forge at Moreton Corbet in Shropshire. Knight's father-in-law appears to have provided the capital for Moreton Corbet and subsequent acquisitions. Knight was running Flaxley blast furnace, Gloucestershire, by 1695–96, and from 1698 the furnace and forge at Bringewood, Herefordshire. To increase his output of pig iron, he built a new furnace at Charlcotte, Shropshire, in 1712 in order to exploit iron ore reserves on the Clee Hills.

The Midland metalware manufacturers benefited from the expanding geography of the trade because they required not only quantities of iron but certain qualities. Iron was not an homogenous product because its strength and malleability varied according to the quality of the raw materials and, to a lesser extent, the quality of workmanship. Forest of Dean ores produced tough iron, the highest grade of wrought iron, noted for its strength and malleability under the hammer. The other main source of English tough iron was the Furness district, which accounts for the early shipment of Furness ore to other ironmaking regions for smelting. Most of the Midland blast furnaces produced coldshort iron from coalfield ores. This made a bar slightly less malleable and liable to shear if it was hammered cold, but it was perfectly suited to everyday items such as nails. By purchasing selected pig iron, Midland forges were able to broaden the product available to manufacturers by offering common and best grades.

The blast furnace and finery forge

Quality was also an important factor in the purchase of imported iron. Swedish and Spanish iron was considered to be of high quality. By the late 1720s substantial cargoes of American pig iron were imported from Pennsylvania, Maryland, Virginia and Carolina. The Crowley and Knight families of the Midland district sold bar iron and rods to the colonial plantations through merchants in Bristol. These same merchants secured imports of American pig iron on return journeys. Imported iron met the shortfall in home-produced iron, but it could also be cheaper. Abraham Spooner started buying Russian iron in the 1730s. At that time Moscow, or 'Mullers', iron was sold at between £11 and £13 per ton for making nails when the equivalent English iron was selling at £18 per ton.

Secondary iron industries

A number of secondary trades were closely associated with the iron industry and it is worth noting them here. By far the most important was the nail trade. Nails were forged from long rods formed in slitting mills. A slitting mill was a type of rolling mill where the roll was fashioned with separate parallel grooves. A bar, hammered or rolled to a wide, flat section was passed through the mill, and was slit into a number of thin rods of square section. The earliest-known British slitting mill was erected in 1590 at Dartford, Kent. The earliest Midland slitting mill was built in around 1611 at Cannock Wood, Staffordshire. Slitting mills proliferated in the Midlands but they were found in all of the ironworking districts. They were comparatively cheap to build and, because the iron was only heated sufficiently for it to be malleable, coal was regularly substituted for charcoal by the eighteenth century.

Wire drawn from iron was introduced into Britain in the sixteenth century. It was an indispensable component of wood cards, and therefore required for one of the most important national industries. William Humfrey, Assay master of the Royal Mint, built

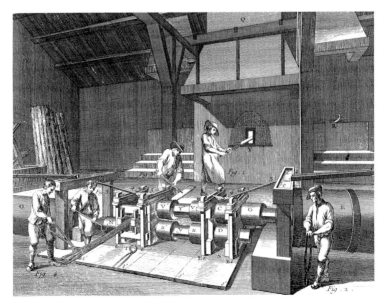

13 An eighteenth-century rolling and slitting mill. Iron is taken from the furnace at the rear of the picture, passed through rollers on the right to form a long flat bar, which is then passed through the slitter to form a bundle of rods.

a wire mill near the ruins of Tintern Abbey in Monmouthshire in 1566–67. Expertise was brought from Germany, first from Christopher Schutz who was principally skilled in brass working, then from Barnes Keysar, a skilled wire drawer. Success led to the erection of a second wire works further upstream on the Wye at Whitebrook in 1607–08. The earliest known description of wire drawing in Britain was made by John Ray in 1675. Iron was hammered into long, thin bars of square section. Its strength was regained by annealing, where the iron was heated up to 900C and was then allowed to cool slowly. Iron was then drawn through a plate with progressively smaller tapering holes, followed by further annealing, the produce progressively finer wire.

In order to reduce iron to a thin section it had to be of the highest quality. Iron for wire manufacture was therefore forged by a different technique to produce 'Osmond' or 'Osborne' iron. This type of iron was produced in a similar way to the finery, except that the semi-molten iron was coiled around the finer's staff in order to form thin layers that ensured more thorough oxidation than was normal. The resulting iron was worked under a lighter and faster hammer than the conventional tilt hammer, and did not need to be re-heated in a chafery. It was quicker, but more physically and mentally demanding work than the conventional finery.

Steel is an alloy with a low carbon content, between ¼ and 1½ per cent. It can be forged and its combined qualities of hardness without brittleness made it particularly important for hardening the edges of tools. Steel was manufactured by a process known as cementation, producing an alloy known as blister steel. Iron bars were packed with charcoal inside a sealed refractory chamber, and were heated for several days to allow carbon to diffuse into the metal. Because the fuel used to heat the iron did not actually come into contact with it, coal could be used instead of charcoal. The process was very sensitive to impurities such as sulphur and phosphorus, and was best made from hematite iron ores which are low in those chemicals. From the 1690s German steelmakers in the Derwent valley introduced the technique of shear steel – the welding of several heated bars of blister steel into a single bar of superior strength – from which time Newcastle-upon-Tyne and the Derwent valley were at the centre of the steel industry until overtaken by Sheffield in the mid-eighteenth century. The Hallamshire district of South Yorkshire, which included Sheffield, specialised in manufacturing cutlery and was the chief centre of production outside of London by the mid-seventeenth century. Allied crafts were the manufacture of edge tools, whose hard steel edges were manufactured from steel shipped from the Weald and Forest of Dean, or else imported from Germany and Spain. Home production began in the second half of the seventeenth century. Subsequent expansion of the steel sector was made possible by the import of high-quality Swedish bar iron, which found an increasing share of the Sheffield market. Henceforth Sheffield was to dominate the British steel industry, more so after an improved technique of manufacture, known as crucible steel, was developed in the 1740s.

The manufacture of frying pans was a small-scale specialisation, but it illustrates a tendency whereby masters of secondary iron trades integrated backwards to forges, and occasionally to blast furnaces as well. Robert Plot visited a pan-making forge at Newcastle-under-Lyme in the mid-seventeenth century. He saw pans hammered into shape in piles, or 'nests', of up to seven plates, of which the lower was the largest. The skill was the ability to hammer all the plates to the required thickness without them becoming welded together. The Newcastle forge was owned by the Holland family, one of the best-documented of specialist pan makers. Cornelius van Halen (born 1581) was a Flemish Protestant who emigrated to England in around 1610 and carried on his trade as a pan maker in Wandsworth near London. Subsequent generations, having adopted the name Holland or Hallen, set up in business in Stourbridge and Newcastle, then branched out. From at least 1660 to 1824 a branch of the family worked at the Lower Forge, Coalbrookdale in Shropshire. In 1753 frying pans, salt pans, plate warmers and dish covers were produced in workshops at the forge and were sold in Shropshire and Wales, or shipped down the River Severn to Bristol. Sons and grandsons of Cornelius van Halen expanded into the finery forge trade in the Midlands. Subsequent generations remained in the forge trade until the early nineteenth century and inter-married with the Wheeler family, an established iron-industry family who had entered the trade as managers for the Foleys.

Nail manufacturers such as John Gibbons (1703–1778) of Kingswinford, Staffordshire, were similarly to found important ironworking dynasties by backwards integration to the forge and then to the smelting sector. So too did many iron merchants, known as ironmongers in the seventeenth and eighteenth century, like the Crowley and Spooner families. One of the most illustrious of ironmonger-turned-ironmasters was William Wood (1671–1730). He began his career at a Wolverhampton ironmongers but by 1715 had formed a partnership for the production and marketing of iron in the Midlands and London. With the grand title of the Company of Ironmasters of Great Britain, Wood's ambition was to transform the partnership into a public company. Disaster struck from an unexpected quarter, as his venture perished in the fallout precipitated by the failure of the South Sea Company in 1720. William Wood was also involved in a failed project to mint copper coinage, one of a number of ironmasters to have interests in non-ferrous metals. One of Wood's partners, Thomas Harvey, managed a new forge built at Tern, Shropshire, in 1710. Described by Harvey as 'the first joint works of its kind in England', it initially combined iron and brass production, incorporating brass rolling mills, and had a cementation furnace for steel manufacture. Iron was not identified by manufacturers as a distinct industry until later in the eighteenth century.

The market for wrought iron

Bar iron was used for a variety of products. There were large-scale producers in the Midlands and Sheffield, serving national and international markets for hardware and cutlery, and small-scale manufacturers of, for example, nails and locks in every market town. Nails were especially important in the building and ship-building trades and therefore were made in most of the ironworking districts. By the early eighteenth century the manufactured iron trades of South Staffordshire and Birmingham included harness wares, scythes and other edge tools, nails, locks, hinges, buckles, toys, swivels, files and chains. Iron merchants sought bar iron from forges in the Midland counties, but also went further afield to Welsh and Derbyshire forges in order to meet the great demand from the manufacturing districts. But there is little evidence that the trade ever grew out of its regional structure. In 1680 an agreement was reached by the Foleys whereby Paul Foley would sell no Forest of Dean bar iron north of Tewkesbury, while John Wheeler and Richard Avenant, managing Philip Foley's Midland forges, would sell no bar iron south of Tewkesbury. It was one of the earliest cartels in the iron trade and acknowledged that ironmasters had regional interests.

Ironmaking

The potential for a national market was inhibited by modes of transport. Carriage over land was cumbersome and slow. Wealden iron was sold to London merchants and was transported by sea from Rye or Newhaven. In the west, the River Severn and its tributaries constituted the most important river navigation system in Britain in the early modern period. The port at Bewdley provided a link to the River Severn for the industrialised areas of the Stour valley and Birmingham. In the late seventeenth century vessels from Bewdley accounted for 20 per cent of voyages through Gloucester, rising to 25 per cent in the early eighteenth century. Downstream traffic was mainly to Bristol and other Bristol Channel ports. Shipments to London from Gloucester or Bristol were not numerous. By 1700 goods could be carried to Lechlade, Gloucestershire, from where the Thames was navigable, but it does not appear to have tempted ironmasters into penetrating the London market. To exploit that market the Stourbridge ironmaster Ambrose Crowley (1658–1713) built forges and slitting mills in the Derwent Valley, County Durham, purchasing most of his pig iron from London merchants. He claimed that transport of nails by sea from Sunderland to London was more efficient than carriage from Birmingham.

The Midland metalware trades were not the sole large-scale market for iron by 1750. The Wealden iron industry was best known for cannon manufacture, for which there was a national market. In the first half of the sixteenth century it had also benefited from the growing trade of London. Bar iron was sold to smiths making nails and other components for the building and ship-building trades, while rural smiths purchased bar iron to make components for ploughs, harrows and wagons. The Weald's market therefore expanded rapidly to encompass the whole of south-east England. Even in the seventeenth century there is little evidence that significant quantities were traded beyond London, or found their way to the Midlands. Competition for the London market in the seventeenth century came primarily from Spain and the Baltic, from where imports of bar iron trebled between 1588 and 1634.

Forges in the north of England also served a largely regional market. John Fell, clerk to Wortley and later to Attercliffe Forge in Sheffield, certainly dealt with London merchants, but most of his trade was with market towns in Derbyshire and Yorkshire, or across the Pennines in Liverpool, Manchester and Stockport. Other Spencer family ironworks dealt with London and Birmingham, but most of the merchants they sold to were in northern towns, like Manchester, Hull, York and Newcastle-upon-Tyne. South Yorkshire specialised in cutlery, but it also had significant allied trades in the manufacture of scissors and shears, scythes and sickles, files, iron boxes, forks and awls. Furness iron was also sold locally, although much of its output was sent to Liverpool for use in the shipyards.

Reinhold Angerstein offers a valuable insight into the iron trade of the early 1750s in his accounts of journeys through Britain. Although not a systematic study of the trade, his wealth of anecdotal evidence illustrates a number of important themes. It emphasises the continued importance of imported iron, despite the rapid growth of the industry in Britain, and that in any given market for bar iron, competition was more likely to come from abroad than from other native ironworking regions. It also highlights the demand for iron of various grades, whose quality could ultimately only be vouchsafed by the customers that used it. Quality was hotly debated. In Barnsley, South Yorkshire, he heard competing claims for the quality of iron drawn from English and Swedish iron. In the market town of Exeter, he found that merchants either purchased iron shipped down the River Severn, or relied on imports. Local blacksmiths' shops made items such as nails, screws for presses, spades, gratings, roasting jacks and fire pokers, but little of its was manufactured from home-produced iron. English iron was cheaper than Spanish and comparable to Swedish prices, but was clearly considered inferior as it tended to be used 'where it does not have to be worked very much and where it is not subjected to heavy stress'.

There is no convincing case that Britain had a national iron industry in the eighteenth century. The potential for a national market was inhibited by modes of transport and therefore the industry developed a strong regional character. The largest of these regional industries was in western Britain, stretching from Cheshire and North Wales to the Wye valley and Forest of Dean, with the River Severn forming its spine. It was in this region that the majority of developments in the eighteenth-century iron industry took place.

4

THE COKE IRON INDUSTRY

Abraham Darby in Bristol and Coalbrookdale

Ironworking at Coalbrookdale probably began in the sixteenth century, when a bloomery is first recorded there. The first blast furnace was built in 1658 and had a number of tenants before Abraham Darby I acquired the lease of it in 1708. The previous tenant was Shadrach Fox, who had abandoned the furnace some four or five years previously. It has been suggested that the furnace was damaged during the floods that followed the Great Storm of 1703, at which point production of cast iron ordnance ceased. The work of the furnace at that time is of particular interest in the light of what happened after Darby assumed control.

14 The forehearth of the blast furnace at Coalbrookdale. The furnace was rebuilt in 1777 but retained lintels of the original furnace that was dated 1658 (the date was changed to 1638 in the 1950s). (© author)

Abraham Darby I (1678–1717) had been a maltster in Birmingham and Bristol before entering the brass and iron trades in Bristol in the first decade of the eighteenth century. Darby had established an iron pot foundry in 1703, buying pig iron from the Forest of Dean. His subsequent move to Coalbrookdale was intended to secure his own supply of pig iron for the foundry. Darby was also a leading partner in a new brass works built at Coalbrookdale at the same time. Darby's memoranda make it plain that, when the blast furnace was repaired and put back into blast, coke and not charcoal was the fuel. Although this has long been interpreted as a pivotal event in the history of the iron industry, and even of the industrial revolution, it is better to look at it as part of a longer process.

In the seventeenth century Dud Dudley (1599–1684) had smelted iron ore with coke but without sustaining commercial production. It was not possible to use coal because it contains sulphur, which contaminated the iron. Coking the coal drove off unwanted impurities. Darby had no previous experience of iron smelting and, just as he had done in Bristol, arriving at Coalbrookdale he must have relied on technical advice, probably from his furnace keeper, who was responsible for managing the charge and ensuring that the furnace yielded pig iron of suitable quality. It is possible that coke had previously been used at Coalbrookdale, if only as a supplement to charcoal, although there is as yet no proof. The pressing need to find suitable fuel in a situation where charcoal was hard to come by may have been relevant before Darby arrived in Shropshire. Darby never attempted to patent his process and there is no evidence that he urged other ironmasters to adopt mineral fuel.

If coke had been used at Coalbrookdale before Abraham Darby, Darby's own difficulties with the material suggest that its use had not been perfected. Despite a promising start, the quality of coke iron was inconsistent, and the quantity of pig iron shipped from Coalbrookdale declined between 1710 and 1712. The chief problem was to find the best grade of coal for conversion to coke, which caused shipments of coal from as far afield as Bristol and Neath to be sent to Coalbrookdale. In the end, the local clod coal, hitherto deemed the poorest of the Shropshire coals, was found to be the most suitable. The major casualty of Darby's period of experimentation appears to have been his brass and copper works. Most of the brass-making equipment was shipped back to Bristol in 1714. The decisive year for smelting with coke was 1715, when a second furnace was built at Coalbrookdale immediately downstream of the original furnace. It was the beginning of a trend whereby fuel was no longer the limiting factor in the number of furnaces at a specific site, but the availability of water power.

Abraham Darby was also influential in developing techniques in the foundry. In Bristol Darby exploited the skills of brass founders by putting them to work casting iron pots in sand moulds. Iron had traditionally been cast into moulds of loam – a mixture of clay, manure, straw and sand – which had to be formed every time a casting was made. With sand moulding, a pattern, usually of wood, was pressed into the sand. Although it was expensive to make, a pattern could be re-used many times. Sand moulds had already been used to cast simple shapes like cast-iron rollers, but casting iron pots in sand was a significant technical advance for which Darby was awarded a patent in 1707. His pots were superior to the crude pots made by the earlier method. For example, five-gallon pots made in 1681 in Shropshire weighed between 37 and 40lb (17–18kg), whereas Darby's pots weighed only 27lb (12kg).

In Bristol, Darby purchased his pig iron from the Forest of Dean and had it re-melted in a reverberatory furnace, which was found to improve the quality of the iron. In a

Ironmaking

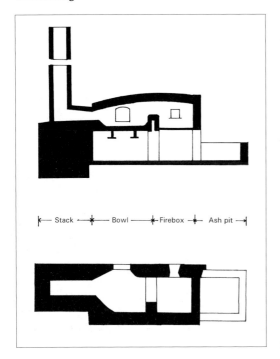

15 Plan and cross-section of an air, or reverberatory furnace.

reverberatory furnace the iron and the fuel were kept in separate chambers. The heat from the firebox was drawn across the bed or bowl, where the metal was placed, by the draught of a tall chimney, obviating the need for bellows. Separation of the fuel from the iron allowed coal to be used because it did not contaminate the iron. The reverberatory, or air, furnace was also relatively new to the iron industry in the early eighteenth century, even though it had already become established in the non-ferrous industries. Previously it had been the custom to cast objects direct from the blast furnace. By using separate furnaces, casting was detached from the management of the blast furnace, allowing greater flexibility in organising production. Where iron was melted in two reverberatory furnaces, much larger items could be cast than had been possible direct from the blast furnace.

By the time of Darby's death at only thirty-nine, Coalbrookdale was unique in the British iron trade because it had two coke-fuelled blast furnaces and reverberatory furnaces that allowed the foundry to work independently of smelting. His success lay mainly on the foundry side of the industry, which was one of its growth sectors but still accounted for only a small proportion of the industry's output. Significantly, the invention of precision casting in the iron industry coincided with the development of the steam engine which, although it is associated mainly with the work of Thomas Newcomen (1664–1729), was covered by a patent awarded in 1698 to Thomas Savery (1650–1715). Early Newcomen prototypes had brass cylinders but the use of cheaper iron cylinders made more economic sense. Abraham Darby's legacy of expertise helped his successors to enter the burgeoning engineering market. The first commercial use of a Newcomen engine was for draining a mine at Dudley and was built in 1712. Seven years later a pumping engine was built for a colliery near Coalbrookdale. The Coalbrookdale works is known to have been involved with casting parts for steam engines from 1718,

The coke iron industry

16 Coalbrookdale in 1758 by Francois Vivares. On the left is an engine cylinder being transported from the nearby boring mill. In the centre is the Coalbrookdale Upper Works, with its blast furnace and two tall chimneys of reverberatory furnaces. On the right is the reservoir for the waterwheels and the coke hearths. In the distance are Dale House and Rosehill House, homes of the ironmasters, and Tea Kettle Row, a row of workmen's cottages. (© Ironbridge Gorge Museum Trust)

and made complete engines by 1722. When Savery's patent expired in 1733 the partners managing the Coalbrookdale works invested in a new mill for boring the internal surfaces of the cylinders, in the expectation of an upturn in orders.

Charcoal iron smelting after 1700

The charcoal sector of the iron industry had not been affected by these developments because it was dominated by the production of wrought iron. In some of the older established ironworking districts, like the Weald and the Forest of Dean, charcoal iron smelting declined in the first half of the eighteenth century. By 1750 Wealden blast furnaces produced less than 10 per cent of national output, compared with over 26 per cent a century earlier. In other areas, such as South Wales and Furness, charcoal iron smelting increased after 1700. The Furness district saw a marked upturn after the first blast furnace was built there in 1694. Furness accounted for nearly 20 per cent of national output of pig iron by 1750, where eight blast furnaces had been built by 1748, including those at Backbarrow (1712), Leighton (1713) and Duddon (1736). Backbarrow was established by Abraham Darby's friend and fellow Quaker, William Rawlinson (and was to survive to the twentieth century as the last furnace in Britain to smelt with vegetable fuel), but he had no interest in converting to coke smelting. When, in 1727, Rawlinson experienced difficulty with the supply of charcoal, his response was

Ironmaking

to ship ore to Invergarry, in Argyll, for smelting instead of buying in coal. Invergarry furnace was short-lived – it was in blast for less than a decade – but it was the earliest Scottish component of the Furness iron industry, combining Furness iron ore with charcoal from the Caledonian Forest. Another blast furnace in the Highlands, at Glen Kinglass, was built in around 1722 by an Irish co-partnership tempted by the timber resources of the Scottish Highlands, but it too was short-lived as it was blown out by around 1738. A second generation of furnaces was built and operated successfully in Argyll at Bonawe and Craleckan, established in 1752 and 1754 respectively. Bonawe (also known as Lorn Furnace) was built near the shore of Loch Etive, not far from the ill-fated Glen Kinglass furnace. Wood rights were granted for a 110-year lease period, which saw charcoal imported from Glen Kinglass woods 5 miles from the furnace, to other sources as far as 40 miles distant. A profitable by-product of the charcoal industry was oak bark, an essential ingredient in tanning, and in its heyday bark was shipped from Bonawe to tanneries on the Clyde and in north Lancashire.

Hematite from Furness was capable of producing wrought iron of high quality. Proximity to the coast at a time when coastal shipping was the principal means of transporting heavy goods has also been cited as a reason why the Furness ironmasters prospered and expanded. The same argument applies to Wales, where most of the charcoal furnaces of the late seventeenth and eighteenth centuries were built within 10 miles of the coast. Abercarn blast furnace was built in 1750, 7 miles north-west of Newport, Monmouthshire. It smelted local ores and hematite from Furness, supplying its own forges at Abercarn, which also had a wire mill, and Caerleon. Pig iron was also sold further afield, including to the Knight family's Stour valley forges. Further inland, Pontypool Ironworks was noted as a pioneer of rolled iron and for its output of high-quality tinplate and wire. Its furnace continued to smelt with charcoal until the end of the eighteenth century.

17 Bonawe blast furnace built in 1753, one of the final generation of charcoal ironworks. (© author)

The coke iron industry

18 Dyfi Ironworks, by J.G. Wood, 1811. Dyfi was one of several charcoal blast furnaces built near the Welsh coast in the mid-eighteenth century. This view shows the blast furnace with charging house behind and casting house to the right. On the right of the picture is the manager's house.

The growth of coke smelting

Only a handful of ironworks had adopted coke smelting by 1750, and then only sporadically. Two other furnaces managed by the Coalbrookdale partners – Abraham Darby II (1711–1763) and Richard Ford (d.1745) – at Willey in Shropshire and Bersham near Wrexham, sometimes smelted with coke. Kemberton furnace in Shropshire was supplied with coke by the same coalmasters who supplied Coalbrookdale. Elsewhere, Bryncoch near Neath, Clifton in Cumbria, Whitehill in County Durham and Sutton in Lancashire are the only other furnaces where documentary evidence has shown the use of coke in the second quarter of the eighteenth century.

The coke iron industry expanded rapidly from the 1750s, due to a number of unrelated factors. The Seven Years' War (1756–63) fuelled the arms trade. It was in the 1750s that coke pig iron was first successfully forged into wrought iron and when the steam engine became widely used in iron smelting. The latter two factors account for the rapid expansion of the industry in the East Shropshire coalfield. Members of his family claimed that Abraham Darby II invented a new process at the forge, but the evidence is unconvincing. Instead, Darby probably discovered that low-phosphorus iron ore yielded pig iron that was acceptable to forge masters and workmen for producing certain types of bar iron. Whatever the case, Darby was quick to seize an opportunity and found financial backing from the Quaker business community in Bristol. In 1754 he acquired a lease of land to the north of Coalbrookdale at Horsehay, and two years later land at neighbouring Ketley. Two new blast furnaces were built at both sites, where Darby ingeniously compensated for the inadequate water supply by using steam pumping engines to recycle water for the waterwheels. It was the technology of the colliery pumping engine adapted to a different use, similar to a scheme that had supplemented the natural water supply for the furnaces and forge at Coalbrookdale since 1743. Significantly, Darby leased his own mineral reserves so that where previously he had relied upon local coalmasters for the supply of coke, he now became a coalmaster himself.

In the East Shropshire coalfield coalmasters responded by building their own blast furnaces. It was therefore an important stage in the merging of two hitherto separate industries. New blast furnaces were built at Madeley Wood, New Willey (both 1757) and Lightmoor (1758). At Madeley Wood, better known as Bedlam, the principal shareholders had been involved in the coal trade before the furnace was built and a lease was agreed that permitted the exploitation of coal and ironstone reserves as well as the erection of the works. It epitomised a new breed of ironworks as an integrated operation that controlled all stages of production from the getting of raw materials to casting the finished product. The engine at Bedlam was used to pump water from the River Severn for its two waterwheels. Other schemes had been used to safeguard the water supply when the natural supply was low, whereas at Bedlam it was impossible for the furnaces to work without it.

Ambitious new coalfield ironworks were also built at Carron (1759) near Falkirk and in South Wales. On the northern rim of the South Wales coalfield all the raw materials needed for iron smelting were easily available. A comparatively isolated district, there was no significant coal trade in the area so the ironmasters had no choice but to organise all aspects of production, as well as transportation of the finished product. A coke-fired blast furnace was begun at Dowlais near Merthyr Tydfil in 1759, and in upland Glamorgan it was soon followed by others at Hirwaun (1757 but originally charcoal-fuelled), Plymouth (1763) and Cyfarthfa (1765), all with rights to dig ironstone and coal.

Although wrought iron was successfully forged from coke pig, its quality was not equal to charcoal pig. Edward Knight (1699–1780), owner of several forges in the Stour Valley, Worcestershire, pioneered the forging of coke pig iron. In 1754–55 he purchased 6 tons of coke pig for Wolverley Forge, rising to 254 tons in the following year, suggesting that the suitability of coke pig iron was rapidly proved. Most of the new Shropshire blast furnaces supplied pig iron to Wolverley Forge in succeeding years. Knight's other Stour Valley forges, at Cookley and Mitton, purchased only small quantities of coke pig iron and concentrated on forging charcoal pig. As late as 1784–85 the Knight family forges purchased 1,729 tons of charcoal pig and 715 tons of coke pig, the latter accounting for only 29 per cent of its iron. Coke pig iron, smelted from coalfield ores, produced the relatively hard coldshort iron, suitable for nail-making but not for high-quality tough iron used in chain-making, wire or tinplate manufacture.

Output of coke pig iron reached 25,000 tons by the mid-1770s, which is probably as much as the charcoal sector of the industry produced at its peak in the 1720s. At the same time charcoal pig iron production was approximately 17,000 tons, but 46,000 tons of bar iron were imported, accounting for 52 per cent of iron used in Britain. Demand for coke pig iron grew steadily in the foundry, partly because there were so many new uses for cast iron. Many of these new uses were pioneered by the ironmasters themselves. By the 1770s cast iron was used for manufacturing waterwheels and flywheels, and the cylinders that replaced bellows for blowing the blast furnaces. Blowing cylinders had been developed by Isaac Wilkinson (1695–1784), firstly in the late 1730s for the forges at Backbarrow, and then at Bersham furnace near Wrexham in the 1750s; cylinders designed by John Smeaton (1724–1792) were used at the Carron blast furnaces from the outset. In 1774 John

Wilkinson (1728–1808), ironmaster of Bersham, New Willey in Shropshire and Bradley near Bilston in Staffordshire, was awarded a patent for an improved method of boring cannon. Ordnance remained lucrative for the iron industry during the war-blighted late eighteenth century, including the American War of Independence and the Revolutionary and Napoleonic Wars with France. The technology could also be adapted to the boring of engine cylinders, a technical lead that allowed Wilkinson to secure contracts from Boulton & Watt for the supply of its patent engines.

There was no better advertisement for the potential of cast iron than the Iron Bridge that spans the River Severn in Shropshire. Designed by the architect Thomas Farnolls Pritchard (1723–1777) and funded by shareholders, the driving force behind the project was the young Abraham Darby III (1750–1789), a third-generation Darby with an awful lot to live up to. The project attracted excitement from the outset, largely due to its vigorous promotion by the shareholders. Erected in 1779 and opened on New Year's Day 1781, it was immediately hailed as one of the wonders of the age, and was an early triumph in the heroic age of civil engineering. In America Thomas Jefferson obtained an engraving and hung it in the White House. Darby received the gold medal of the Society of Arts in 1787, his compensation for what turned out to be a financial disaster. Numerous cast iron bridges were subsequently built and there were other innovations in cast iron and engineering. An iron bridge at Cyfarthfa, Merthyr Tydfil, was built in 1793 that incorporated an iron water channel. Soon the technique was applied on a larger scale, notably the iron aqueduct that was built on the Shrewsbury Canal at Longdon-on-Tern in 1796. In the same year an iron-framed mill was erected at Ditherington in Shrewsbury, although iron columns and beams had already been incorporated into other mills and churches.

19 Calcutts Ironworks on the south bank of the River Severn, Shropshire, in 1788. Calcutts specialised in casting cannon during the lucrative, war-torn years of the late eighteenth and early nineteenth centuries. (© Ironbridge Gorge Museum Trust)

Ironmaking

20 The Iron Bridge, by Michael Angelo Rooker, 1782, dedicated to King George III and one of the most popular contemporary views of the bridge. (© Ironbridge Gorge Museum Trust)

21 The towpath and water channel of Longdon-on-Tern aqueduct on the Shrewsbury Canal, built in 1795–96. It was one of the first generation of iron aqueducts. (© author)

New coal-based techniques in the forge

In 1750 the wrought-iron sector of the industry still relied on charcoal. During the latter half of the eighteenth century numerous attempts were made to replace it with mineral fuel. Where fuel was required simply to bring the iron to a welding heat, for example in the chafery hearths and slitting mills, coal had often been used as a cheaper alternative. Several new techniques using mineral fuel were patented between 1761 and Henry Cort's puddling and rolling process, patented in 1783 and 1784. Apart from their use of coal or coke, these new techniques also marked the appearance of the reverberatory furnace in the forge. The simple fact that iron and coal were kept in separate chambers made reverberatory furnaces attractive to innovators, because the chief problem with coal in the refining stage was sulphur contamination.

Charles Wood (1702–1774), of Low Mill forge in Cumbria, and his brother John Wood, of Wednesbury forge in Staffordshire, obtained patents in 1761 and 1763 for a technique known as stamping and potting. Sons of the ironmaster William Wood, the brothers had been brought up with the trade and inherited their father's interest in developing new processes. Wood senior had wanted to smelt iron ore in a reverberatory furnace to yield malleable iron, and in 1753 Charles Wood remained optimistic that it would be possible. Reinhold Angerstein visited Low Mill in 1753 and observed another of Charles Wood's experiments, in the reworking of scrap iron. Old nails, locks and keys, mainly imported from Holland, were cleaned of rust and then heated inside clay pots in a reverberatory furnace, where the iron coalesced and formed a solid mass that could be drawn into bars under a hammer. Wood evidently worked on this process for some time. By partially refining pig iron in a finery using coal or coke, and then breaking the resulting iron into small particles (stamping), he could place the granulated iron inside clay pots (potting) for the reverberatory furnace. The process was evidently commercially successful. It was used by the Woods at Low Mill and Wednesbury. When Charles Wood's business partner and brother-in-law, William Brownrigg, became a partner in the proposed new Cyfarthfa Ironworks in Merthyr Tydfil in 1765, Wood was sent to Wales to supervise the erection of the furnace and stamping-and-potting forge. Cyfarthfa thus became one of the first ironworks of the coke era designed to integrate furnace and forge using mineral fuel alone.

Stamping and potting competed with other processes patented in the 1760s, none of which were commercially successful, but all of which highlight the importance of the endeavour. Carron was an ambitious enterprise that intended to integrate smelting, founding and forging at a single site. From here, John Roebuck (1718–1794) was awarded a patent in 1763 for refining pig iron in a traditional finery and chafery, both using coal or coke. Two Coalbrookdale workmen, Thomas (1711–1780) and George (born 1701) Cranage, were awarded a patent in 1766 for refining iron in a reverberatory furnace, with financial backing provided by their employer, Richard Reynolds. Like stamping and potting, the technique was derived from techniques of reworking scrap iron, but equal success was not achieved using pig iron.

John Wright and his brother-in-law Richard Jesson (1741–1810) were forge masters and among the most prominent nail makers and ironmongers in West Bromwich, Staffordshire. In 1773 they were awarded a patent that improved the Woods' original stamping-and-potting process. Their business continued after the death of John Wright in the 1770s, who was succeeded his son Richard. A second patent, awarded in 1784, specified that the iron could be heated on stacks of clay tiles rather than in pots, a technique known as piling. There are three detailed accounts of Wright and Jesson's version of stamping and potting, two written at Coalbrookdale in 1785 and 1803, and the earlier written at Wright and Jesson's West Bromwich forge in 1775.

Marchant de la Houliere, ironmaster and erstwhile brigadier in the French army, was despatched to Britain by the French government to report on English industry. At West Bromwich he persuaded the workmen to demonstrate their new process, probably by offering the appropriate fee. The pig iron was refined in four stages. Iron, mixed with scale and cinder, was heated in a traditional finery using coal. Four pigs, weighing approximately 1cwt (51kg) each, made up a single charge. The heated iron, now in the form of spongy balls, was removed, laid on a flat cast iron plate, and hammered to a flat cake about 1½in thick. Once it had cooled it was placed back under the hammer and broken up into small pieces (stamping). The granulated iron was washed to remove extraneous material, including coal cinders, and was then left to dry. In the final stage the iron was placed in pots 12in high and 11in diameter, and placed in a reverberatory furnace (potting). The final stage lasted between four and five hours, after which the pots broke up and the iron particles inside each pot had coalesced. These were hammered individually and then sent to a chafery, from where they were hammered into bars ready for sale.

Stamping and potting was successfully employed at several Midland ironworks in the final quarter of the eighteenth century – by 1790 at a minimum of nine forges in Staffordshire and eight in Shropshire. In some places it was introduced at new ironworks where smelting and forging were integrated from the outset, such as Donnington Wood (1785) and Old Park (1790) in Shropshire. At others, such as Bradley in Staffordshire, Horsehay and Ketley in Shropshire, stamping-and-potting forges were built alongside existing furnaces and represented an expansion in the range of operations. Stamping and potting was also introduced at what had been finery forges. Wright and Jesson's West Bromwich forge, and the forge they leased at Wren's Nest in Shropshire in 1776, are good examples. Overall, the adoption of stamping and potting in the Midlands heralded a drift towards the coalfields in the industry's centre of gravity, and saw coalfield ironmasters expand their interest from pig and cast iron to include wrought iron.

The drift to the coalfield was also aided by the parallel development of steam technology, which in the eighteenth century was mainly confined to small pockets of the South Wales and Midland iron industries. In the 1760s James Watt (1736–1819) had improved the original Newcomen engine by designing a separate cylinder for condensing the steam, greatly increasing its efficiency. After entering into partnership with Matthew Boulton (1728–1809) in Birmingham orders for its patent engines started to grow. Boulton & Watt made only small components and provided specific designs. For the larger components they worked closely with ironmasters who had the expertise to cast precision items, chiefly John Wilkinson. Previously engines had worked to raise water, either from mines or from the tail races of waterwheels where water was to be recycled. New innovations included blowing engines, whereby the engine drove a blowing cylinder for the blast furnaces, replacing the waterwheel altogether. The first blowing engine was constructed at John Wilkinson's New Willey ironworks in 1776, and a total of sixteen had been built by 1800, with a further two adapted from pumping engines. John Wilkinson was both the principal supplier and the biggest customer, supplying blowing engines to each of his works at Bradley in Staffordshire, New Willey and Snedshill in Shropshire, and Brymbo in Denbighshire. The next stage was to design an engine that could work a forge hammer, although Watt had to find an alternative to the crank shaft, which was already covered by another patent. Instead he designed 'sun-and-planet' gear, employing it for a hammer engine at Bradley in 1782, and for engines at the three Coalbrookdale Co. forges – Horsehay, Ketley and Coalbrookdale – when they adopted stamping and potting in 1784–85.

Stamping and potting and steam power each contributed to a significant increase in output. British bar iron output rose from 18,800 tons in 1750 to 32,000 tons in 1788. Individual forges were also more productive. A survey of the forge trade in 1749 found that the most productive of the nation's finery forges, Mitton and Wilden in Worcestershire, each produced 450 tons of iron at two hearths, although the national average was only 211 tons per forge. In 1794 two hearths at Horsehay produced 694 tons of iron to be worked into bars.

There is no evidence that rural forges or the finery and chafery were threatened by the emergence of stamping and potting. A viable coal-based method of refining pig

The coke iron industry

22 The Upper Forge at Coalbrookdale in 1789. Originally a finery forge it was converted to a stamping-and-potting forge in around 1783, and investment was made in steam engines in 1785 and 1787. (© Ironbridge Gorge Museum Trust)

iron merely added to the complexity of the wrought-iron trade. The new coalfield blast furnaces in Shropshire, and to a certain extent in Staffordshire, produced more pig iron than their associated forges could refine. Consequently they relied on small independent forges to purchase their surfeit of iron, most of which were charcoal fuelled. Evidence also shows that many coalfield ironmasters operated both coal and charcoal technologies at different sites. Thomas, William and Benjamin Gibbons owned stamping and potting forges at Lye, Cradley and Level in Staffordshire, but continued to operate Pitchford finery forge in Shropshire. Likewise Samuel Hallen owned Wednesbury Forge in Staffordshire, where stamping and potting was introduced in 1786, but was also a partner in Tibberton and Upton finery forges in Shropshire. Thomas Botfield (1736–1801) built the last of the ironworks integrating smelting with stamping and potting, at Old Park in Shropshire in 1790, but retained Cleobury Dale forge in south Shropshire to work with charcoal.

Technological advances of the eighteenth century – coke blast furnaces, steam engines, stamping-and-potting forges – all combined to create a new type of ironworks by the 1780s. Ironworks had become self-sufficient in the acquisition of iron ore and coal, and integrated multiple blast furnaces and a forge on a single site. It was an operational structure that would become even more dominant in the nineteenth century when puddling and rolling had become the most successful method of manufacturing wrought iron on a large scale.

5

PUDDLING

Cort, Crawshay and the development of puddling

Henry Cort (1740–1800) was a Navy Agent in Portsmouth, responsible for the disbursement of monies to Royal Navy ships and the Royal Dockyards. The dockyards consumed immense quantities of wrought iron for anchors, bolts, spikes, nails and other uses. This was a potentially lucrative market for British ironmasters, if only they could persuade the Navy Board to accept iron from home producers, instead of purchasing what it considered to be a superior product from Sweden. Ironmasters were not slow to point out the strategic danger of relying upon foreign iron. Malachy Poslethwayt wrote in 1747 that, '*Bar Iron* is one of the *Essentials* belonging to our *Naval Stores*; and a happy Independency upon any foreign Nation for our supplies which so nearly concern the Support of our *maritime Power*, has always been judged desireable'.

In the 1770s Henry Cort took over the management of a forge at Fontley near Portsmouth from his uncle, William Attwick, and concentrated upon winning contracts to supply the dockyards. When, in 1780, the Victualling Board awarded him a contract to supply barrel hoops, he entered into a partnership with Samuel Jellicoe. Jellicoe's father, Adam, was Deputy Paymaster of the Royal Navy and advanced them money from the balance of public funds he held, a practice not unusual for the time. Cort then set about trying to compete with Swedish iron with a home-produced version of equal quality. The result was puddling, for which Cort was awarded patents in 1783 and 1784. Although Cort's motive was commercial, its success depended upon garnering a wide repertoire of manual skills in the forge and developing them into something new.

His first patent was for the working up of scrap or bad iron (scull iron) that had already been refined. Both the common finery and the reverberatory furnace were specified. Iron could be worked in several ways. For example, scull iron could be piled up on old plates in a reverberatory furnace; alternatively old salt-pan plates and other plates were cut up, formed into coffin shapes and filled with scraps of nut and bracket iron. Enclosing the small iron in a makeshift container was similar to heating granulated iron in pots, just as placing small iron on plates was derived from piling. The radical element was Cort's claim that by passing the iron through a rolling mill a superior quality of iron could be produced. The rolling mill was not new in the iron trade, but hitherto it had been specialist plant used for rolling plates and, by the late eighteenth century, for rolling boiler plates for steam engines. Cort claimed that 'mooring-chain links, ships' knees, and other iron decayed and eaten by rust, being cut into proper lengths, duly heated and passed through the rollers, will produce exceedingly good iron without any other process'. The rolling mill produced a more consistent size and shape of bar than the forge hammer, but it also aided the removal of slag, which had previously been achieved by forgemen entirely under the shingling hammer.

The principal element of his second patent was the refining of pig iron in a reverberatory furnace. Pig iron, to which small amounts of fragmented scull or scrap iron were added, was brought to a white heat in the furnace. A blue flame (carbon monoxide) signalled to the forgeman that the carbon in the iron was reacting with oxygen, which was accompanied by a vigorous stirring of the metal to allow maximum exposure of the iron to oxygen. Although the pig iron had previously begun to melt, when the iron came to its pure state it melted at a higher temperature. The iron therefore formed into a spongy mass that the forgeman removed from the furnace for shingling under a hammer. It was then placed in another or the same furnace, and then passed at white heat through a rolling mill, the effect of which was that 'the iron will be discharged of the impurities and foreign matter that adheres to them when manufactured in the methods commonly practised'.

The essential characteristic of puddling was its reliance upon manual dexterity. It was a technique, not technology. This is a crucial point because it sets innovation in the forge sector apart from contemporary developments in the smelting sector like the invention of steam blowing engines. Whereas an engine's motion was predictable and repeatable, the product of a puddling furnace depended upon a variety of factors, not least the skill and judgement of the workman.

Cort sought to persuade ironmasters to take up the process. Demonstrations were made at various ironworks in the Midlands, for example at Shut End and Hyde in Staffordshire, before heading north to Carron in Scotland. Ironmasters and workmen were unimpressed with Cort's claims, none more so than the workmen at Ketley in Shropshire. The ironmaster William Reynolds had kept a close watch on the development of new processes for the forge. Early in 1784 he had allowed Peter Onions (*c*.1720−96) to conduct trials at Ketley of a process similar to puddling, and was apparently disappointed that they did not succeed. And it was after Cort's trials that he made a significant investment in stamping and potting.

The Ketley workmen claimed that puddling was derived from the Cranage brothers' process patented in 1766. When Cort required written confirmation that his workmen had demonstrated the process to Ketley workmen, it was signed by Thomas Cranage and Thomas Jones. Thomas Cranage was the son of the patentee George Cranage while Thomas Jones been the executor of the will of the patentee Thomas Cranage, who had died in 1780. Cort found himself up against men who wanted no lessons in how to work iron in a forge, and his visit became the stuff of myth. An apocryphal story, said to have derived from Joseph Reynolds (1768–1859), younger half-brother of William and eventual successor as owner of the Ketley Works, claimed that Thomas Cranage bested Henry Cort by proving that his breakthrough was nothing new. Apparently Cranage 'put in some white iron − cold-blast mine iron − and soon brought out a ball of puddled iron'.

The story illuminates the degree of pride that was at stake. Given that Cort's men were working iron in a reverberatory furnace just as the Cranage brothers had done, to achieve a superior bar iron from Ketley pig iron could only mean that Cort and his men had developed a superior technique. Forgemen with an obvious local loyalty are unlikely to have conceded the point. In fact, Cort was unable to persuade an iron industry largely composed of men bred up in the trade that he, a relative newcomer, could work iron with more dexterity and economy than they. Cort failed to licence his process at any coalfield ironworks. After the first round of public trials the only ironworks that agreed to take up the process and pay a royalty was the Rotherhithe Co. of London.

Undeterred, Cort continued to advocate the advantages of his new process. Trials were made at various dockyards – Deptford, Chatham, Woolwich and Plymouth – to test the comparative strength of anchors, bolts and tackle hooks manufactured from Swedish and from puddled iron. In each case Cort claimed that puddled iron was equal to or better than its Swedish counterpart. The aim of these trials was to persuade the Navy Board to purchase puddled iron and to stipulate that iron sold to the dockyards should be refined in this way. The effort appeared to have paid off by 1789, when the Navy Board stipulated that iron should be made to Cort's patent when inviting tenders for a new contract. In Richard Crawshay, Cort had also found an ironmaster willing to invest in his new process.

Richard Crawshay (1739–1810) was the most successful London iron merchant of his day, having been sole proprietor of a merchant house since 1763. He amassed his fortune importing Swedish, Russian and German iron, making him acutely aware of the potential for the British iron industry if it could produce iron of equal quality. In 1777, after the start of the American War of Independence, he had become a partner in a cannon foundry at Cyfarthfa Ironworks, Merthyr Tydfil, having obtained a contract with the Board of Ordnance for the supply of munitions. However, not until 1787 did Richard Crawshay become the leading partner at Cyfarthfa, following the death of Anthony Bacon, its founding partner. By now a wealthy man but with undiminished ambition, Crawshay saw an opportunity to create the most modern ironworks in Britain and to become the nation's leading ironmaster. Within months Crawshay became the second ironmaster to take out a licence for the puddling process, agreeing to pay Cort a royalty of 10s for every ton of iron produced.

Crawshay's optimism was soon dented. Iron was puddled at Cyfarthfa and sent to Rotherhithe for rolling, but the quality was poor. Cort had sent Fontley workmen to Cyfarthfa to instruct them in the new process, but the workmen evidently found it difficult to master. Crawshay vented his frustration on Cyfarthfa's managing partner, James Cockshutt, nephew of a Yorkshire ironmaster and someone with a long experience in the trade. Cockshutt had apparently been exhorting his forgemen to shorten the refining time as a way of decreasing costs, but it proved a false economy. If the iron was inadequately refined it was of uneven quality, part of it being malleable but with other parts brittle (known as red-short) under the forge hammer. When poorly refined iron was passed through the rollers at Rotherhithe an unacceptable proportion of the iron peeled off as mill scale.

The situation was remedied partly when Crawshay sent his son William to Merthyr to direct the erection of a rolling mill in 1789, with technical supervision from Henry Cort. Then, unexpectedly, disaster struck. Cort's ironworking business was based upon loans amounting to about £27,000 made out of public funds. When the Treasurer of the Royal Navy called in the debt, Cort and Samuel Jellicoe were unable to pay. Bankruptcy and the confiscation of Cort's patent rights followed.

Although Cort remained in contact and ostensibly was on amicable terms with Crawshay, he soon faded from the scene. Cockshutt was dismissed in 1791 and in 1792 Crawshay moved to Merthyr to manage the ironworks himself, leaving the charge of

23 The head of the Glamorganshire Canal at Cyfarthfa Ironworks, by William Pamplin, c.1800. The house is Cyfarthfa House, home of Richard Crawshay, and to its right is the puddling forge erected under the supervision of Henry Cort. (© Cyfarthfa Castle Museum & Art Gallery, Merthyr Tydfil)

his merchant house to his son William. By now, Crawshay's investment in puddling had begun to pay off. At the end of 1791 he claimed to have 'at last overcome the Evils of pudling', the first known use of a term never used by Henry Cort. Modifications were made to the construction of the reverberatory furnace by replacing the clay roof with a cast iron plate. The bed of the furnace was made up of clean sand. Mostly, however, the problems were about technique. Cyfarthfa retained its stamping and potting forge and the practice of granulating refined iron was still valued. Writing in 1791 Crawshay described the process then in use:

> We make use of Air Furnaces instead of Finerys, when the metal is brot into nature, instead of Hammers, we put it between a pair of Rolls, & crush it like a paste about ¾ in thick – then break it into small pieces, pile up to 60 to 80 lbs wt on a Cake of baked Clay, heat 20 of those piles at a time in [another] Air Furnace, then shingle them under a Hammer of 1200 lb weight fixed in an Iron helve of 2 Tons wt. The Blooms thus finished is again heated in an Air Furnace & brot into a very handsome Bar by Groov'd Rollers.

The most significant improvement on Cort's original specification concerned the quality of the iron charged into the furnace. Since 1788 the Penydarren Ironworks, established in Merthyr Tydfil in 1785, had been supplying puddled iron to Crawshay's merchant house in London. Here it was found more efficacious if the pig iron was first re-melted in a refinery, also known as a running-out fire. This was a simple melting furnace, blown with tuyeres on either side, from which molten iron was run out to form a flat plate. Once solidified, the iron could be broken up for charging into the puddling furnace. The process decreased the silicon content of the iron, with a consequence that, once cooled, what had been grey pig iron changed into white refined iron. The running-out fire was never patented and its use appears to have spread quickly.

Ironmaking

24 Penydarren Ironworks, Merthyr Tydfil, by J.G. Wood, 1811. Penydarren was the second ironworks to adopt puddling successfully. The puddling forge is in the foreground, beyond which are two rows of workmen's cottages and, in the background, the blast furnaces.

The growth of the South Wales iron industry

Crawshay claimed to have invested £50,000 at Cyfarthfa between 1787 and 1793, much of which was on increasing the scale of the works. By 1791 Cyfarthfa already had blast furnaces 60ft (18.3m) high – nearly twice the height of contemporary Shropshire furnaces – each capable of producing 1,400 tons per annum. Its forges had an even greater capacity, having produced 2,300 tons of bar iron in 1790 by buying pig iron from neighbouring ironworks. In 1794 there were two blast furnaces; a third was added in 1796 and a fourth in 1797. Output had risen to 6,000 tons by 1798, making it by far the largest ironworks in Britain. Two blast furnaces were built on a new site at Ynysfach in 1801.

Cyfarthfa was well placed for large-scale iron production, an advantage it shared with other ironworks at the heads of the South Wales valleys, providing ironmasters could find sufficient capital to exploit their potential. At Penydarren, profits made in the 1790s were ploughed back into the business, increasing the capital value of the works from £14,000 in 1786 to £46,000 a decade later. Dowlais, starved of investment in its early years and where puddling was adopted comparatively late in 1800, the capital value rose from £8,000 in 1786 to £120,000 in 1804.

The first generation of coke ironworks in South Wales were let on favourable terms by landlords who could not foresee how lucrative their mineral reserves would prove to be. The 1765 Cyfarthfa lease gave it mineral rights across a 4,000 acre estate for a term of ninety-nine years and for an annual ground rent of only £100. The neighbouring Dowlais works was even luckier, having secured in a lease negotiated in 1749 mineral rights over 2,000 acres of common land for a ground rent of only £31. Landlords

Puddling

25 Cyfarthfa Ironworks by J.G. Wood, 1811. On the left are the blast furnaces and casting houses, which have distinctive round vents in the gable. On the right are the puddling forge and rolling mills, and a row of workmen's cottages behind.

26 Rolling mills at Cyfarthfa Ironworks by Penry Williams, 1825. Viewed at night to increase its dramatic effect, reverberatory furnaces can be seen on the right, in which bars were heated ready for the rolling mills on the left. The crane was used to lift and replace components of the rolling mills. Note the presence of boys among the men in the mill. (© Cyfarthfa Castle Museum & Art Gallery, Merthyr Tydfil)

Ironmaking

quickly changed their tune when ironmasters started making handsome profits. In 1789 Blaenavon was let on a lease of twenty-one years at an annual rent of £1,300, although it included an enormous 12,000 acre tract of land, so large that a separate ironworks was built at Nantyglo on part of the same leasehold in 1792. When the lease was renewed in 1806 the annual rent was increased to £5,200. This should be compared with ironworks in other districts. For example, in 1781 the Coalbrookdale Co. leased works and minerals for the Coalbrookdale and Madeley Wood Ironworks for a ground rent of £2,000, in a district where minerals had been worked for over a century. Furthermore, in Shropshire the raw materials could be won only below the surface. At the heads of the South Wales valleys, coal and ironstone were more abundant and could be dug from the surface.

'We can find materials for six more Furnaces if we could find Money to build them', claimed William Lewis of the Dowlais works in 1790. Plenty of other investors could see the potential of South Wales, where ironworks were established at Sirhowy (1778), Beaufort (1780), Blaenavon (1789), Ebbw Vale (1789), Nantyglo (1792), Clydach (1793) and Neath Abbey (1793) that exploited the increasing demand for iron from the 1780s. The success of puddling provided an even greater incentive, with new works built at Tredegar (1800), Union (1800, and later extended on a new site and known as Rhymney Ironworks) and Aberdare (1802), in addition to the expansion of existing works.

This generation of ironworks profited from the success of steam technology, without which its development would have been inhibited. Neath Abbey, Tredegar and Union Ironworks were all established with Boulton & Watt blowing engines (the latter a second-hand engine from Shropshire), and similar blowing engines were bought at the end of the 1790s for established works at Penydarren and Sirhowy. At Penydarren

27 Blaenavon Ironworks in 1801, by Richard Colt-Hoare. Blaenavon was built in 1789 and was one of the largest of the many ironworks built along the heads of the valleys in upland Monmouthshire and Glamorgan.

investment was also made in an engine to drive the rolling and slitting mill. Gilbert Gilpin's survey of Monmouthshire ironworks in 1796 also describes blowing engines at Blaenavon, Ebbw Vale and Nantyglo.

Significant advances in steam technology should not be interpreted as signalling the immediate demise of the waterwheel. Water power remained important to the iron industry, not simply because it was cheaper or because ironmasters were too conservative to invest in steam engines. Where the water supply was good it was a perfectly satisfactory source of power. Archaeological and documentary evidence has revealed the continuing importance of the waterwheel into the nineteenth century. At Coalbrookdale, both blast furnaces relied on water power until smelting ceased in about 1817. The turning mill at the upper works, for smoothing the surfaces of rolls, cylinders and pipes, remained water-powered for the remainder of the nineteenth century. The nearby Bedlam blast furnaces were blown by means of a waterwheel until smelting ceased in 1843. In South Wales, Blaenavon had been built relying completely on steam power for its furnaces, but when a new forge was built on the hillside at Garnddyrys in 1816 it was water powered. Excavation at nearby Clydach Ironworks, although it did not cover the entire area of the works, revealed the original wheelpit built in 1793, but it did not yield the expected evidence of a later engine house, and may therefore have remained water powered either wholly or partly well into the nineteenth century. Watkin George, chief engineer at Cyfarthfa Ironworks, built in 1796 a waterwheel, nicknamed *Aeolus*, to blow four blast furnaces. The wheel was 50ft (15.2m) in diameter and 7ft (2.1m) wide, driving two flywheels and four blowing cylinders 44 inches (1.1m) in diameter. The wheel became something of a spectacle in its own right, and continued to blow the furnaces until

28 Clydach Ironworks by J.G. Wood, 1811. Behind the furnaces figures can be seen working in the coke yard, beneath the raised water launder supplying the blast furnace waterwheel.

Ironmaking

29 Bont Fawr aqueduct spanning the Afan valley upstream from Port Talbot, built in the period 1824–26 for Oakwood Ironworks. (© author)

steam power replaced it in the 1820s, by which time the number of furnaces had already increased to six. The Bont Fawr aqueduct, built to supply waterwheels at Oakwood Ironworks in west Glamorgan, which began production in 1826, remains a magnificent monument spanning the Afan valley, and testament to the local climate.

Puddling was successful in South Wales, partly because it was introduced into a region where the new ironworks did not have a long-established forge culture. As it had been pioneered in South Wales it became known colloquially as the 'Welsh Method' (although 'Merthyr method' would have been more accurate). At one time historians believed that the confiscation of Cort's patent rights precipitated the widespread adoption of puddling because a premium would no longer be payable for its use. Although the evidence of Cyfarthfa shows this not to have been the case, the impression has still been left that the 1790s saw a revolution in the iron trade. In reality the pace of change was not so quick, nor the future direction of the industry so definite as it seems in hindsight. By 1805 South Wales accounted for 30 per cent of pig iron smelted in Britain, by which time Shropshire's share of the market had fallen from its high of 40 per cent to only 22 per cent. Even so, smaller, older forges continued to operate in South Wales. The wire and tinplate industries continued to purchase charcoal-refined wrought iron from small forges like Pontypool, and Dulais Forge near Neath. Other small forges adopted puddling, like Glangwryne near Abergavenny, which had three puddling furnaces by 1796, although the iron was still hammered rather than rolled.

Forges in England in the early nineteenth century

The adoption of puddling in the Midlands lacked the initial high investment seen in South Wales, partly because ironmasters had recently invested significant amounts in the stamping-and-potting process. Technological change was consequently slower and more piecemeal. At John Wilkinson's Bradley Ironworks, Joshua Gilpin described in 1796 a works that had fully implemented Cort's specification of puddling furnaces and rolling mills, together with the preliminary refineries. In Shropshire the situation was different, and illuminates how new technology was introduced in an already established coal-based iron industry. A useful comparison can be drawn between Samuel and Jeremiah Homfray's Penydarren Ironworks and their brother Francis Homfray's various works in Shropshire and Staffordshire. At Penydarren, the blast furnaces and puddling forge were integrated on a single site. Francis Homfray adopted puddling at Lightmoor, the site of his Shropshire blast furnaces, where iron blooms were formed under the hammer. These were sent to his other forges, at Hyde and Swindon in Staffordshire, both of which had rolling and slitting mills, for working into sheets, nail rods and bars. Therefore, although Francis Homfray embraced new technology, the conveyance of semi-finished iron between forges over a long distance continued a long-established pattern of working wrought iron.

In 1794 Old Park became the first Shropshire ironworks to invest in puddling, followed by Ketley by 1796. Ketley was owned by William Reynolds, one of Richard Crawshay's closest friends in the iron trade. The two men exchanged information on inclined planes and iron aqueducts and Crawshay had long urged Reynolds to invest in puddling. When Joshua Gilpin visited Ketley in 1796 he saw running-out fires and puddling furnaces. However, instead of passing the iron through a rolling mill it was rolled into flat cakes,

30 A schematic view of rolling at Horsehay Ironworks, Shropshire. On the right are a puddled ball and a shingling hammer. The mills on the left include mill trains for bar and sheet. In the foreground, the catcher can be seen passing the bar back over the top of the rolls for the rollerman to make another pass.

granulated and heated inside clay pots before being formed into blooms under the hammer. These were then sent to the Coalbrookdale Co.'s Horsehay forge for rolling into bars. Evidence from Old Park suggests the same partial adoption of the new technology, with no immediate decline of stamping and potting. Investment in steam-powered rolling mills was made at Old Park in 1801 and at Ketley at about the same time, but stamping and potting remained valuable for better-quality iron. In particular it was used for the manufacture of boiler plates, which had been one of the key products of Horsehay rolling mill from the 1780s, and was used for reworking scrap iron at Horsehay until 1830. Reworking scrap iron was not as insignificant as it might sound. In the period 1802–03 it accounted for 5–10 per cent of the Horsehay forge output.

In the Midlands, small independent forges continued to characterise the wrought-iron industry. Even where investment was made in puddling by the larger smelting concerns, it was insufficient to refine all the pig iron smelted in the furnaces. At Old Park, for example, 5,500 tons of pig iron were smelted in 1815 but only just over 2,300 tons were converted to wrought iron in the company's own forge. The remainder was sold to independent small forges, most of them in the Midlands. The Knight family's Stour valley forges demonstrate how these small forges came to concentrate upon special grades of iron for which there had been a long-established market. The Mitton Upper and Lower Forges were converted for stamping and potting as late as 1796–98. The Upper Forge also had puddling furnaces, blooms from which were sent to the Lower Forge for stamping and potting, with other outlying forges used for drawing, slitting into rods for nails, and rolling. This pattern of working continued until 1814, when Knight closed the Upper Forge and converted the Lower Forge to a charcoal finery.

The success of puddling did not precipitate the demise of charcoal ironmaking. In the Midlands many finery forges did close in the early nineteenth century, but in a competitive market those with poor transport links could not survive. Well-placed forges like Wilden and Broadwaters close to the Stour in Worcestershire continued to exploit the market for high-quality iron.

The long overlap between stamping and potting and of puddling and rolling demonstrates that technology was not an either/or option but the gradual incorporation of new techniques alongside established practices. The forge trade continued to hold in high regard the value of granulating semi-finished iron and re-heating it in pots or piles. Stamping and potting was therefore valuable beyond the self-contained process patented by the Wood brothers and Wright and Jesson. Seen in isolation, the sequence of patent processes delivers a linear model of technological change in the eighteenth century that does injustice to the complexities of the industry. Each process codified a range of techniques that were utilised by forgemen in various combinations to produce iron of specific qualities. Survival of what has been deemed obsolescent technology was governed by the eventual use of the iron, be it charcoal iron for chain links or stamped iron for boiler plates.

The iron trade at work

Although puddling has often been interpreted as revolutionising the iron trade, a brief survey of the market for iron and the manner in which the iron industry did its business will help to put the technology in context. Until the railway boom in the 1830s, only a small proportion of iron was sold direct to customers. Precision engineering items, mostly the product of foundries, were the largest category. Most iron was sold in semi-finished form to other ironworks for further refining, or to iron merchants, many of whom, such as those of London, Bristol or Liverpool, purchased iron destined for export.

The iron industry needed a market place which, by the second half of the eighteenth century, was served by quarterly or more regular meetings of ironmasters and merchants. These were regionally based, the earliest and largest of them being the Midland quarterly meeting. As an institution, the Midland quarterly meeting appears to have evolved from regular gatherings at fairs in Stourbridge, Bristol and Chester, where outstanding accounts were settled. The Stourbridge meeting was established before there were significant numbers of blast furnaces in the Black Country, but by the early nineteenth century the quarterly meeting was held in Birmingham. As early as 1731 Edward Knight, owner of several Stour valley forges, was already talking of setting prices on a quarterly basis. The tradition that the price of iron was decided by the largest manufacturer persisted into the nineteenth century. The Midland quarterly meeting was the best organised of regional groups, and survived until its functions were taken over by the British Iron Trade Association in the 1870s. Other areas were less formally cohesive. The Yorkshire and Derbyshire ironmasters set up a local association only in 1799, but it fizzled out in the 1820s and the only issue it successfully debated was the price of iron.

Quarter days were Lady Day, Midsummer, Michaelmas and generally two weeks after Christmas. Ironmasters used the opportunity to request payments and as the principal forum for receiving orders. The rationale was explained by William Botfield (1766–1850), managing partner of Old Park Ironworks: 'Our method of doing business is we generally take at quarter day orders for nearly the whole make of our iron, to be delivered in the course of the quarter which we do as regular to each [merchant] house as we can'. This allowed an ironworks to organise its production over a three-month period. It also informed ironmasters what grades and types of iron were in demand, and presented choices to them of either producing a variety of special products, or risking production on fewer, more regular types of bar iron in the hope that the price would remain profitable. By taking orders at quarterly intervals, ironmasters could ensure that no iron was left unsold. Business was organised in such a way that the finished product was in circulation between producer and consumer for as short a time as possible. Short-term strategies were adopted to cope with price fluctuations where stockpiling would have been a longer-term solution. Ironmasters were not always in headlong pursuit of economies of scale. Gilbert Gilpin, clerk to the Old Park ironworks, explained in 1808 that flooding the market was counter-productive:

> iron would soon feel a greater depression in price from the augmentation of make still more exceeding the consumption. It is on this account that we confine our mill ... to eight or nine hours work per day, and roll 50 tons of bar iron per week. In lieu of selling large quantities at prime cost, my employers conceive it best to do less at a small profit.

Quarter days were also partly a social occasion. They ensured that an ironmaster had a fairly good idea of what his competitors within the region were doing. Such contacts were also useful when ironmasters needed to unite in defence of their interests. A succession of measures that proposed to levy increased taxes on iron and coal production, in 1785, 1796 and 1806, were successfully defended by ironmasters who could draw benefit

Ironmaking

from well-organised district or regional societies, as well as its network of partnerships linked by inter-marriage. They provided a forum for shop talk but not, apparently, for the dissemination of technical knowledge. In 1800, Joseph Dawson, managing partner of the Low Moor ironworks in Yorkshire, attempted to introduce a formal discussion of metallurgical science at the meetings of the Yorkshire and Derbyshire ironmasters, but to no avail. No one but himself was prepared to read a paper on the subject.

Prices of iron and coal, and wages for iron workmen and colliers, were fixed at quarterly meetings. Since few of the major ironworks kept significant stocks of iron, it was permanently in their interest to keep the price of iron at a profitable level. Quarterly meetings facilitated this as they were a way of checking the pulse of the trade. Those ironmasters, like the Crawshay and Gibbons families, who bestrode both the manufacturing and mercantile sectors of the trade, could find themselves in an advantageous position during depressed times. When the price fell ironmasters responded by producing less while the Crawshays could stockpile Cyfarthfa iron in London and wait until the price rose again. Prices and wages were established by custom but no one was obliged to stick to them. Ironmasters could simultaneously lower their price to attract custom from their competitors and complain that their rivals had increased wages, making them attractive to valuable skilled workmen. Ironmasters co-existed in a tension between unity and competition.

The successful and widespread adoption of the rolling mill in the wake of puddling brought a wide range of sections of iron on to the market. The choice faced by ironmasters

31 Rolling mills at Dowlais in the late nineteenth century. On the left is the reverberatory furnace with its tall stack, and on the right the mill trains and a crane for lifting and replacing the machinery. The picture is dateable to the late nineteenth century by the presence of men rather than boys as helpers. (© Ironbridge Gorge Museum Trust)

was how many of these shapes should the mill produce, and to what sizes. An example of the proliferation of sections is provided by the price list of the Dowlais Ironworks in 1816. It includes thirty-three different sizes and sections, including eight sizes of rounds and six sizes of flats. The Old Park rolling mill had separate trains for different types of iron – bar of large and small sections, plate, sheet, hoop and rod – that were coupled up only when they were in use. Rolling mill engines could not work all the mill trains simultaneously and so organising the efficient use of the mill to fulfil its quarterly orders was a fine calculation. Some of the smaller sections required at least two heats because the iron could not be passed through the rolls fast enough before the iron had cooled. The cost of manufacturing wrought iron therefore depended upon the size of the bar sold to the customer. Regulations were agreed that only one tenth of an order for bar iron could be of those deemed to be small sizes. Gilbert Gilpin complained to one customer that:

> were we to manufacture a greater proportion of [small sizes] at the prices mentioned in the list, we should sustain a loss of at least £4 per ton upon the excess, arising from extra waste of iron, additional expense of workmanship, and their employing our machinery five times as long as the same weight of assorted sizes.

In 1808 small sizes included square bars of ½in, flat iron of between 1 and 1¼in thickness, and round iron between ⅜ and ¾in diameter. The variable price of bar iron is not therefore a reliable guide to the profitability of the refining process, and nor can the profitability or importance of a forge be measured simply as a statement of output in tons.

Dialogue between masters and merchants depended for its technical vocabulary on the language of workmen. The arcane glossary of terms to define aspects of furnace and forge was part of the industry mystique and helped to foster a conservative, inward-looking culture. The use of technical vocabulary in contemporary documentary sources is always at one remove from its source, since these documents were written by managers and ironmasters. Moreover, as an oral culture the language was never definitively codified. For these reasons the use of terms can seem idiosyncratic and are not always susceptible to definitive interpretation. Meanings also changed over time and so a term must be interpreted within its specific context.

The emergence of the term *puddling* in the last decade of the eighteenth century has already been mentioned, but it did not refer exclusively to Cort's patent, as it could also refer to the reworking of scrap iron in a reverberatory furnace. Similarly the term *balling* was a general term applied to the re-heating of semi-finished iron. Although it commonly referred to the re-heating of puddled iron, the term might have originated in the re-heating of stamped iron inside pots (the term *potting* was rarely used). By the end of the eighteenth century finery forges also had balling furnaces, either to replace the chafery or to work scrap iron, and so the term encompassed both charcoal and coal technology. Heating stamped iron on clay tiles was known as *piling* in the contemporary trade, as was used by Richard Crawshay describing the forge at Cyfarthfa in 1791, but it differed from the subsequent meaning of the term. Later in the nineteenth century *piling* was a common term for re-heating stacks of bars in a balling furnace.

Pig iron self-evidently belongs to the introduction of the blast furnace, and remained in use in the coke era. *Bloom* and *shingling* also survived the transfer to coal technology. Bloom is of medieval origin and was always used to define iron once had been completely refined, or 'come to nature' in late eighteenth-century forge-speak. Other terms faded from common usage. *Coldshort* and *tough* iron were familiar grades of wrought iron in the eighteenth century, but disappeared in the nineteenth century. At Horsehay, in the 1790s, stamped iron was graded *good*, *bad* and *second best*, which is similar to terms such as *good iron* and *better iron* used by Wednesbury nailers in 1785 in trials with puddled blooms. By the nineteenth century iron was graded as various degrees of *common* and *best* iron. Pig iron was graded *No.1* and *No.2* for foundry iron, and *No.3* for forge pig iron, previously known as *grey forge pig*. Ad hoc changes in grading were not always universally understood or accepted. In 1826 two Bristol iron merchants ordered *No.2* and *No.3* bar iron from Old Park in Shropshire, which prompted a swift reply: 'We beg to observe we do not know what numbers mean for bar iron, always distinguishing ours by Common, Best & Best Best, and shall be obliged by your so describing it in future'.

Nothing illustrates the conservative tendency of the iron trade better than its concurrent employment of two standards of measurement. Both shortweight tons of 2,240lb and longweight tons of 2,400lb were in use. In Shropshire, forgemen were paid piece rates for longweight tons for most, but not all processes. Sales of iron in the market were governed by custom. Pig iron, blooms and slabs were generally sold in longweight tons, while bar iron and plate were sold shortweight. In other words, semi-finished iron slabs sent to a mill for rolling into boiler plates were weighed in longweight, but when the plates were returned to the customer they were accounted for in shortweight.

The topic that dominated communication between ironworks and their customers was the quality of iron. In cases where a large quantity of business correspondence has survived, as for example with the Dowlais and Old Park ironworks, complaints abound, to the extent that disputes about quality were part and parcel of business life. Ironmasters and their agents needed a thick skin. In some disputes the ironworks were clearly at fault. In 1822 the London merchants Thomas & William Hood returned to Dowlais iron bars 'so remarkably bad' that they had been rejected by three of their customers as intolerably brittle. 'Please inform us what you wish done with it. We should advise your shipping it to India, as we are sure it would bring your iron into disrepute were it to be sold here.' In 1849 a consignment of Dowlais rails was criticised for having excessive numbers of crooked rails and '125 rails full of cracks and flaws'.

High quality of wrought iron was determined either by the quality of the pig iron, additional manufacturing processes or a combination of both. Blending of pig iron from different sources was a common way of manufacturing superior grades of bar iron. Blending was also recommended for using pig iron at the foundry. In 1784 William Taitt at Dowlais informed customers that 'the iron made at this furnace is not only very soft, but some of the strongest in the kingdom & will bear a large proportion of inferior iron to be mixt with it'. Quality was not a scientifically verifiable standard, but a judgement on the part of the furnace keeper. Pig iron was broken and assessed according to its degree of greyness, its grain and its hardness. Complaints about the quality of pig iron could therefore be explained as careless selection at the furnace.

Ultimately, the quality of iron could only be judged by the customers who used it. The later the stage of working, the more scope there was for poor workmanship accounting for the loss of quality. Blooms and bars were generally not marked by individual workmen, so in the purchase of semi-finished iron it was difficult to pinpoint a cause or a person for loss of quality. Puddlers and shinglers were not disposed to accept responsibility for bad work.

Grading of iron was not successful unless customers bought the right iron for their specific needs. Again there was plenty of scope for disagreement between supplier and customer, and ironmasters had a vested interest in persuading their customers to purchase more profitable high-quality iron. James Sheward, traveller on behalf of the Botfields' Old Park Ironworks, was told in 1815 to encourage customers to take at least a sample of best iron: 'if they will only take half a ton it will give them the opportunity of deciding which answers best for their purposes, for many Ironmongers are quite ignorant of the difference in quality and only judge by the lowness of price'. Price was an unfair criterion. Inferior grades of wrought iron were harder to work, became brittle when worked to a thin section, and when passed through a rolling mill more of the iron was lost as mill scale than with best-quality iron. 'In most instances [the lowest priced iron] will prove much dearer in the end than the best'. However, quality control had little effect without the ability to mark the iron clearly. Steel stamps were used to stamp bar iron according to source and quality, in addition to which paint was used to distinguish best grades and sizes too small to mark in any other way. There was scope for incorrectly stamped iron, and even greater scope for mixing up qualities in transit.

Ironworks despatched their iron to an agreed destination, beyond which the customer was responsible for transport to its final destination. The Midland trade was dominated by the River Severn, where iron was traditionally priced to Bewdley, followed from the 1770s to Stourport. Canals such as the East Shropshire Canal and Staffordshire and Worcestershire Canal linked the ironmaking centres with the arterial route. All iron was weighed at its delivery point along the river, the weights being transmitted back to the source of iron and the invoice drawn up on that basis. (Less iron arrived at the weighing point than was despatched from the works because a certain proportion was stolen en route.) The Severn carriers therefore had an important role as arbiters of goods and were at a pivotal position between supplier and customer. Once iron had been delivered to the carriers' wharves, it was they who had responsibility for delivery to its ultimate destination. Well-established rules governing the Severn trade, and the fact that ironmasters despatched their iron immediately, meant that there was little scope for exploiting competition between carriers. The Severn trade caused difficulties that ironmasters could do little to remedy, although evidence for mixed cargoes suggest that even in the 1820s not all iron bars were effectively marked. Mixing up orders, late delivery, no delivery and the storage of iron in the open air where it gathered rust were common complaints. When deliveries fell short of the amount specified on the invoice, the carriers were responsible for reimbursement of the customer. In 1819 James Robertson & Co. of Glasgow complained that they had received none of the consignment of iron shipped from Old Park. It was conveyed first by river to Bristol, to where it could be accounted for, but was then lost in transit between Bristol and the Clyde. The dispute was settled by an independent arbitrator, who in this instance found in favour of the ironworks because it had fulfilled its contract to deliver the iron to Bristol.

It has been argued that the effectiveness of the River Severn as a means of transportation diminished from the late eighteenth century, in inverse proportion to the demands placed upon it by the ironmasters. For example, proposals put forward in 1786 to improve the navigation were not carried out. But the river remained a cheap mode of transport before the railway age and it should not be assumed that the canal network serving other regions was not just a different type of impediment.

Ironmaking

The profitability of the industry enabled the South Wales ironmasters to overcome the main obstacle to their business, that of transporting iron from a comparatively isolated upland district. The Merthyr ironmasters all invested in the Glamorganshire Canal, which opened from Merthyr to Cardiff in 1794. The decision to build the canal was one of the few collective agreements ever made by the ironmasters. Subsequent relations were characterised by argument and dissension. Richard Crawshay was the main shareholder and dominant on the canal's executive committee. He ensured that the head of the canal was extended to his own works (*see fig. 23, p.49*). It rankled with his neighbours – Guest of Dowlais, Hill of Plymouth and the Homfrays of Penydarren – who in 1802 built a 9-mile tramroad parallel to the upper section of the canal to decrease canal tolls and bypass the congestion. Without the canal Merthyr would never have achieved its prominent position in the nation's iron industry. In 1796 8,370 tons were carried on the canal, rising to 119,853 tons by 1835. Increased traffic brought severe congestion, alleviated only partly by dredging a deeper channel to accommodate narrow boats of larger carrying capacity.

Other South Wales ironworks, mostly situated along the northern rim of the South Wales coalfield, were in no less isolated positions. Monmouthshire ironworks were served by two linked canals, the Monmouthshire Canal of 1792–99 and the Brecknock and Abergavenny Canal of 1796–1812. The position of the ironworks, such as Nantyglo and Blaenavon, meant building horse-drawn tramroads from the works to the various canal basins. Blaenavon was initially served by a six-mile tramroad to the Monmouthshire Canal. By 1817 it had been superseded by a shorter tramroad to the Brecknock and Abergavenny Canal, but over mountainous terrain that necessitated driving a tunnel over

32 Baileys warehouse on the Brecknock and Abergavenny Canal, built in 1821. Iron was brought here by tramroad from the Bailey Brothers' Nantyglo Ironworks. (Crown copyright: RCAHMW)

a mile long, and building a series of inclined planes to convey the iron down the steep slope of the Blorenge.

Reliance upon a few arterial inland routes made transportation a significant part of an ironmaster's business. It was one of several factors that contributed to a well-established, conservative culture in the iron trade. By 1830 the iron trade conducted its business in much the same way that it had done half a century earlier, despite the shift in the focus of the industry towards the major coalfields. The customs dictating traffic on the River Severn are a reminder that old-established rules were not always of the iron industry's making. The conservative approach to business practice also serves to broaden the focus and to check the over-emphasis placed upon technology in the history of the iron industry. To the ironmaster, technology was a tool in the management of the business. The iron trade was driven less by technological developments than by the provision of specialist products in an environment influenced by custom as well as economics. Ironmasters had an ingrained short-term outlook. Their principal focus was upon gaining orders for their quarterly make of iron and to secure an acceptable and realistic level of profits from their investments. Profit was not always achieved by maximising production. The interpretation of the industry in terms of output therefore gives only a limited understanding of the trade.

6

THE NINETEENTH CENTURY

The growth of the iron industry

The iron industry was prone to periods of peaks and depression, a state brought about by the ingrained short-term business strategies of the ironmasters and the volatile market price for iron. The euphoria of Waterloo and the ensuing peace was followed by one of the biggest slumps suffered by the iron trade, resulting in unemployment and social unrest throughout the British iron industry. In South Staffordshire in 1816 only thirty-five of its seventy-one blast furnaces were in blast. But it was only a temporary setback. Despite the vicissitudes of the market, the iron industry continued to expand at an accelerating rate in the nineteenth century. In 1791 output of British blast furnaces exceeded 100,000 tons, a figure doubled a decade later, and doubled again to 400,000 tons by 1810. Output dipped to 260,000 tons in 1817 but had reached 700,000 tons by the late 1820s and over 1 million tons by 1837. Output in 1870 was over 5.9 million tons. South Wales and the English Midlands continued to be the largest of the ironworking regions in the first half of the nineteenth century, although they were surpassed after 1850 by the growth of ironmaking in the north of England and Scotland, whose presence accounts for such vast increases in output during that period. Other important factors in the growth of the industry included improvements in technology, particularly in the smelting sector, and the creation of new markets, particularly the railway industry.

The largest profits from the iron industry had been made in South Wales, where a stream of speculative ventures was initiated in the second quarter of the nineteenth century. New works in Monmouthshire included Coalbrookvale (1821), Blaina (1824), Pentwyn (1825) that later merged with Golynos (1837), Victoria (1837) and Cwm Celyn (1839), and in Glamorgan Gadlys (1828) and Llynfi (1839). In west Glamorgan a patent was awarded in 1836 to George Crane of Ynyscedwyn Ironworks for smelting iron with anthracite, as opposed to the bituminous coal found elsewhere in South Wales. It stimulated further investment in the smelting sector, although anthracite was never successfully employed in the puddling process. Ystalyfera (1839), Wenallt (1839), Onllwyn (1844) and Banwen (1848) were new ironworks hoping to exploit the discovery, although success was patchy.

Not all of these new ventures in South Wales were wise investments, especially when profits were taken as a foregone conclusion. The British Iron Co.'s blast furnaces were built at Abersychan near Pontypool in 1826–27. Abersychan was the first joint-stock speculation in the iron industry following the repeal of the Bubble Act in 1825.

The nineteenth century

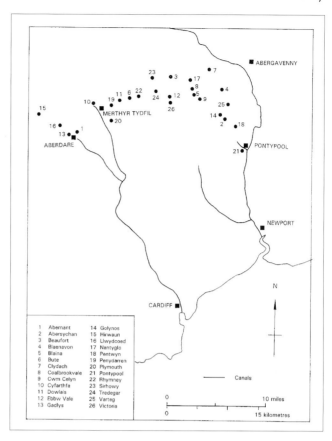

33 Ironworks in south-east Wales in the mid-nineteenth century.

Its cast houses and engine house were designed by the London architect Decimus Burton, forming an impressive and regular classical façade, an impression of authority, order and stability, but also tinged with hubris. It also hid what was by all accounts an ironworks badly built and poorly planned, where the retaining wall behind the furnaces collapsed before the furnaces were completed. It was also established on a disadvantageous lease that helped to ensure that no profits were made until 1838, even though the works became an established producer. The British Iron Co. collapsed in 1843. It successor, the New British Iron Co., went bankrupt in 1851. Valued at £324,003 in 1841, Abersychan was sold a decade later to the Ebbw Vale Co. for a mere £8,000.

An even more architecturally ambitious scheme was the Bute Ironworks, built near Rhymney in Glamorgan from 1824, and designed by Dr James Macculloch (1773–1835). The initiative for a new ironworks had been taken by the landowner, the second Marquess of Bute (1793–1848), who persuaded William Forman, principal partner in the neighbouring Rhymney Ironworks, to invest as the main partner in the new venture. The project was distinguished by its misplaced grandeur. Macculloch's original plan to have twenty-four blast furnaces was fantasy. Forman himself made unrealistic plans for twelve furnaces. In the event three blast furnaces were built in a distinctly Egyptian style, with battered sides, and a casting house with colonnades of lotus columns reminiscent of temples like Dendera. For the engine house, housing a beam blowing engine built at Neath Abbey, the façade adopted the pylon motif and was loosely based on the Egyptian Hall at Piccadilly, London. The boiler house stack resembled a minaret. Given the enormous wealth and power enjoyed by the ironmasters, a comparison between South Wales and the land of the Pharaohs was not inappropriate. The works was troubled in its early years, largely because Forman could not meet his financial obligations and expended most of his energy in trying to extricate himself from the project. The Marquess of Bute complained in 1831 that 'nothing could be more wasteful and absurd than much of the expenditure at the Bute Works'. Few people seem to have been impressed by Macculloch's exotic fantasy. Instead, Bute Ironworks was said to be 'the ridicule of the whole neighbourhood'. Although Bute Ironworks failed to live up to expectations, it was not a complete failure. It merged with the Rhymney Ironworks in 1837, by which time it also had a forge and, when new furnaces were built in the late 1830s, they were of a more conventional design.

In the Midlands, the smelting branch of the industry had gathered pace ever since the introduction of beam blowing engines. In 1788 there were only six blast furnaces in South Staffordshire, but by 1815 there were fifty-five blast furnaces with an output of over 100,000 tons of pig iron. Output of pig iron in the South Staffordshire Black Country peaked in 1852, when 725,000 tons of iron were smelted. North Staffordshire had only two blast furnaces at the end of the eighteenth century. Like the Black Country, North Staffordshire was well endowed with the necessary raw materials and its iron industry expanded in the second quarter of the nineteenth century, aided by the arrival of the railways and the improved transport links it offered. Earl Granville's three furnaces at Etruria started production in 1841, and he had eight furnaces by 1860, making 50,000 tons of pig iron annually. New smelting works were built in the small East Shropshire Coalfield, supplying both the foundry and forge trades of the Midlands. Older furnaces near the River Severn, at the south end of the coalfield, closed down, but the deeper mineral deposits on the north side of the coalfield stimulated many new blast furnace sites. The Lilleshall Co. emerged as the most prominent local producer, superseding the Botfields just as they had in turn superseded the Coalbrookdale partners. In the early 1830s the Lilleshall Co. had blast furnaces at Donnington Wood and Lodge, the former superseded by a new site at Priorslee in 1851, and an engineering works was built in 1861. A forge at Snedshill was operated as a semi-independent concern from c.1830.

The mid-nineteenth century also saw the opening up of new ironworking districts across Britain. In particular, the invention of the hot blast was to have a marked impact on the geography of the British iron trade. It was patented by James Neilson (1792–1865) of Glasgow in 1828. It was originally intended to be applied to cupola furnaces and forge hearths, but its effectiveness when applied to the blast furnace was soon apparent. There was some initial resistance throughout the industry, not least because royalties were payable for its use until 1842. There were also technical concerns. Hitherto the conventional wisdom had been that the colder the blast of air, the more efficient the furnace, on the basis that the yield of blast furnaces was better over the winter months than during the summer. In practice hot blast was quickly proved to be a cheaper way of smelting because it achieved significant savings in fuel, and allowed coal to be mixed with coke in the blast furnace. By the middle of the century it had been widely adopted in established ironworking regions, where cold-blast smelting was now reserved for special grades, usually of foundry iron. Its first great impact was in Scotland, where the rise in output of pig iron was coupled with the discovery of rich deposits of blackband iron ore. In 1830 Scottish blast furnaces smelted 37,500 tons of pig iron

The nineteenth century

34 A blast furnace at Darlaston in the Black Country, showings its lifts, hot-blast pipes and boiler stacks.

35 Earl Granville's blast furnaces at Etruria in north Staffordshire. The shape of the blowing engine house on the left indicates that it housed a beam engine. The blast furnaces are characteristic of the period – masonry bases with arched openings, supporting round brick stacks clad in iron sheets. Also typical of Midland ironworks, the furnaces were built on level ground with raw materials raised to the charging level by means of a hydraulic or pneumatic lift.

67

Ironmaking

36 Blists Hill furnaces, by Warington Smyth, 1847. The blast furnaces and engine houses were built in 1832 and 1840. (© Ironbridge Gorge Museum Trust)

37 The Lilleshall Co.'s Lodge Ironworks in Shropshire, which specialised in cold-blast smelting. Behind the furnaces are three calcining kilns, charged by means of wagons hauled up a short incline on the left of the picture.

from twenty-four blast furnaces, output that rose to 195,000 tons from fifty-four blast furnaces by 1839.

The rapid development of smelting in Cleveland and north-east England occurred after 1850 for similar reasons, exploiting cheap local iron ores and no longer inhibited by relatively modest coal reserves. New dominant figures emerged, including Isaac Lowthian Bell (1816–1904), partner in Clarence Ironworks, Middlesborough and Walker Ironworks in Newcastle-upon-Tyne, and the partnership of Thomas Bolckow (1806–1878), a German merchant, and John Vaughan (1799–1868), an engineer trained at Dowlais. Bolckow, Vaughan & Co. began smelting at Witton Park, near Bishop Auckland, in 1846, and at Eston near Middlesborough in 1852. The Consett Iron Co. of County Durham was formed in 1864 after the collapse of the Derwent Iron Co., and had twelve blast furnaces by 1873. During this period the region became Britain's leading producer of pig iron – Cleveland alone had 104 blast furnaces in 1872 and produced more than 1.9 million tons of pig iron – and the refining sector followed. By 1873 several ironworks – Consett in County Durham, Tees Side and Britannia Ironworks in Middlesborough, Darlington Iron Co., and Witton Park – had more than 100 puddling furnaces each. Characteristically the ironworks of the north east were established as companies with limited liability, made possible by the Companies Act of 1862.

New technology in the iron industry

Although the growth of the iron industry depended upon exploitation of new mineral reserves and building more furnaces, it also benefited from the improved productivity of blast furnaces. Blast furnace capacity was steadily increased. Unlike older furnaces that were constructed around a solid masonry casing, usually against a bank, furnaces from the mid-nineteenth century were provisional structures on cast iron or stone bases, with tall brick stacks clad in iron sheets. From the 1830s various experiments were undertaken to utilise the waste heat from the blast furnaces, applying it to reduce the fuel required for the steam engine boilers. Such experiments followed pioneering work by John Urpeth Rastrick (1780–1856) who, from 1827, had attempted to re-use the heat of puddling furnaces. Where furnaces were blown with hot blast, the air was heated to 200–320C in chambers known as stoves or ovens. In early hot blast stoves air passed through cast iron pipes in brick-built ovens, whereby the heat was allowed to circulate around the entire diameter of the pipe. In the second half of the nineteenth century the most common form was the Cowper stove, first employed in 1860 at Ormesby near Middlesborough and the invention of E.A. Cowper (1819–1893). The stoves comprised tall shafts of checker-work brick in which air was allowed to flow over a large surface area. The stoves worked on a regenerative principle. Hot gases from the blast furnace were drawn through one stove in order to heat the brick. Then air for the blast was blown through the stove while the waste heat from the furnace was directed through a second stove.

Improved steam-engine technology also aided the growth in size and output of individual works. For many years after it had been superseded for other uses, high-pressure beam engines remained the preferred option in ironworks, especially when reliability was the most important consideration. High-pressure vertical and horizontal-mounted engines driving crank shafts could also be put to work blowing furnaces or

Ironmaking

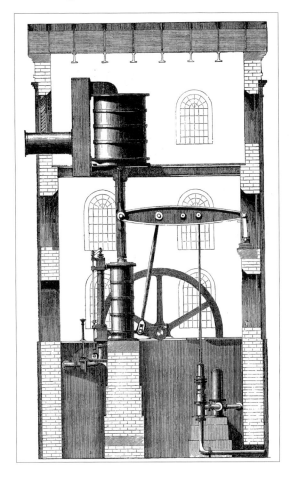

38 A vertical-mounted blowing engine at Barrow Ironworks, built in 1857, by Thomas Perry & Sons of Bolton.

powering rolling mills. In 1863 Ebbw Vale had four beam engines driving the rolling mills, three smaller horizontal engines for driving blooming rolls, two horizontal engines for straightening rails, and two other engines of unspecified type for sawing the ends of rails. Two years later a new beam blowing engine was installed, which had a massive 36ft (11m) beam, a steam cylinder 12ft high (3.7m) and 6ft (1.8m) in diameter, and a blowing cylinder with a 12ft (3.7m) diameter.

Minor improvements were made to the puddling process in the first quarter of the nineteenth century. The first successful use of cast iron bottoms in puddling furnaces has been attributed to Samuel Rogers of Nantyglo Ironworks, Monmouthshire, in 1818. It contributed to the success of pig boiling, the principal innovation in the forge sector during this period. Pig boiling was developed from about 1816 by Joseph Hall (1789–1862) of Tipton, Staffordshire. It is also sometimes referred to as 'wet puddling'. Instead of sand, iron-rich slag was used to line, or 'fettle', the puddling furnace, and being rich in iron oxide it caused a violent reaction with the iron charged into the furnace, with the result that the contents of the furnace appeared to boil. After initial success, problems were encountered because the iron-rich cinder was found to erode the cast iron bottom of the furnace, not solved until the furnace was lined with roasted cinder, known as bull-dog. A patent for roasting the cinder was awarded in 1828.

The nineteenth century

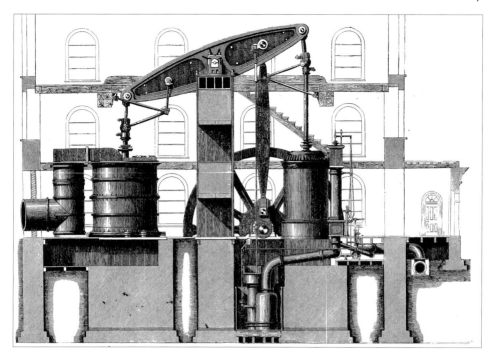

39 A beam blowing engine at Ebbw Vale Ironworks, by the Perran Foundry, Cornwall. Power cylinder and flywheel are to the right, and the pair of blowing cylinders to the left.

Other forms of technology also influenced the forge in the second half of the nineteenth century. Steam hammers came into widespread use after 1850 to replace traditional tilt hammers (see p.8). The earliest was the invention of James Nasmyth (1808–1890), introduced in 1842, followed by several other variants by engineers such as John Condie. The overlap between new and traditional technology was a long one. One of the initial obstacles to wholesale adoption of the steam hammer was that it was more complicated than the simple tilt hammer. Parts for tilt hammers, as for puddling and other furnaces, could be obtained easily from local foundries. The steam hammer was a more complex machine requiring more specialised components and skills in maintenance, and was supplied by engineering works.

There was also resistance to steam hammers on the part of shinglers and perhaps also conservative ironmasters. As it fell by gravity, the tilt hammer provided blows of constant force, which enabled the shingler to judge the quality of iron according to its malleability and how much iron was lost during the hammering. By contrast, the force of a blow could be controlled when using a steam hammer, and an insinuation developed that it could successfully shingle inferior iron where a tilt hammer could not, with a consequent loss of final quality. The disadvantage of the tilt hammer was that the power of the blow decreased as the size of the ball placed under it increased, for the simple reason that the head fell a shorter distance. The steam hammer, on the other hand, could deliver a heavy blow whatever the size of ball placed under it. This made it particularly suitable where iron of large scale or section was being forged, for example, rails, engineering sections and ship plates, but less attractive for the smaller-scale sections used by the hardware trades. Round Oak Ironworks at Brierley Hill, Staffordshire, was

Ironmaking

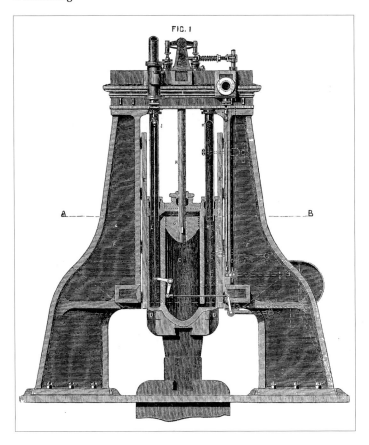

40 A steam hammer at Atlas Works, Glasgow..

the most modern plant in the Midlands when it was built for the Earl of Dudley in 1855, but the tilt hammer was preferred to the steam hammer. At its closure in 1894, the Congreave Ironworks at Cradley Heath, Staffordshire, operated both steam and tilt hammers.

Another advantage of the steam hammer was that a single hammer could cope with the output of approximately thirty puddling furnaces. Coupled with the building of steam engines of greater power, it allowed more powerful and larger mills to be erected. Three-high mills were invented in the early nineteenth century and allowed the iron to be passed through the mill from either direction, eliminating the need for a dead pass. A further innovation was the invention of continuous mills. Charles While, of the Taff Vale Works in Pontypridd, Glamorgan, patented in 1861 a continuous mill whereby iron was guided through consecutive pairs of rolls to produce bars or blooms of the required thickness, thereby bypassing much of the manual labour required for the old method. An alternative was the reversing mill, pioneered by John Ramsbottom at LNWR's Crewe works in 1866. The mill was geared in such a way that it reversed every few seconds and was especially suitable for heavy work. A continuous mill was also devised in 1861 by Charles Bedson of the Manchester Wire Works for drawing wire rods. In practice British wire producers did not adopt the new technology in significant numbers until the last decade of the nineteenth century, after the technology had become established in Germany and the USA.

The nineteenth century

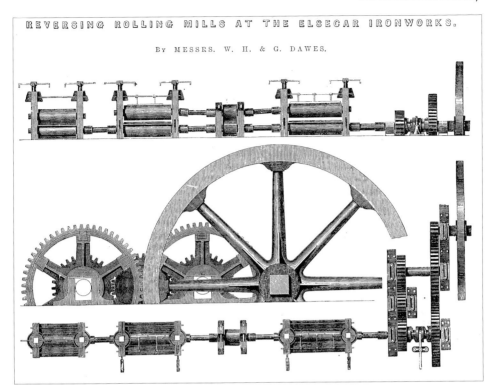

41 Reversing rolling mills at Elsecar Ironworks.

Ironmaking in the mid-nineteenth century

The workings of the iron industry in the mid-nineteenth century can be reconstructed from several contemporary sources. The blast furnaces were operated from two levels – an upper level for charging the furnaces and a lower level for tapping them – and for that reason it remained advantageous to built blast furnaces against a natural bank. At the upper level there was an area for materials preparation known as the coke yard and mine yard. It was usually here that the coal was converted to coke and that the ores were calcined. The other element of the charge, the limestone used as a flux, was used untreated, but it had to be broken down so that each lump was no larger than the size of an apple.

Raw coal was converted to coke either by burning it in open heaps, or in coke ovens. Although coke ovens gradually superseded the open heaps, it is not helpful to see a linear progression from one to the other. Local preference appears to have been an equal factor determining the method used. To burn coal in open heaps two general methods were employed. Coal was placed on the ground loosely enough to allow air to pass through it, and was built up in a heap about 5ft (1.5m) high. In the Midlands the most common method was to build a small round turret of bricks, up to 2ft (0.6m) in diameter, with gaps to allow the air to flow, and then the coal was heaped up around it. On top were layers of poorer slack coal and watered coke dust, known as blacking, heaped up to the

Ironmaking

42 Coke hearths in the early twentieth century. Coal was heaped around the brick turrets, which acted as a draught while the coal was slowly burned. (© Ironbridge Gorge Museum Trust)

shape of a Bronze Age barrow between 18ft (5.5m) and 30ft (9.1m) diameter. The heap was ignited by throwing live coals down the turret, allowing the fire to spread gradually. The length of firing varied according to the size of the heap – which could amount to between 13 and 20 tons – and the quality of the coal. In contemporary accounts firing varies between three and twelve days. The coal burned slowly and gave off large amounts of smoke, although if the wind whipped through the heap it would burst suddenly into flame. To dampen the flames a covering of earth and ashes, or straw and tree bark, was also added. When the firing was finished the heap was doused with water. A large heap of 20 tons would yield about 13 tons of coke.

Before burning, ironstone was laid out on the mine banks in order that it could lose some of its moisture. It was then either burned in open heaps or in calcining kilns. Again, local preference played a large part in determining which of the methods was used. A calcining kiln was similar to a lime kiln in its construction and operation. Coal and ironstone were charged in layers from the top, and after it was sufficiently roasted, the iron was 'drawn' from the bottom. To calcine it in open heaps, coal was laid on the ground, above which the ironstone was laid loosely in a heap, in order to allow the circulation of air. Then further alternate layers of coal and ironstone were added to create a mound which was covered in coal dust to inhibit the escape of air and therefore to allow the minerals to burn slowly. In contemporary accounts the length of calcining varied enormously, from a few hours to twelve days. Optimum size of a piece of ironstone for the blast furnace was about the size of a man's fist, requiring larger lumps of ironstone to be broken down.

In the eighteenth and early nineteenth centuries it was commonplace for a charging house (or bridge house) to be built at the top of the furnace. Its existence was a throwback to the days of charcoal blast furnaces, when it was imperative to keep the fuel dry. The charging house was invariably a warm place where there were only intermittent

The nineteenth century

43 Ironstone mines at Madeley Wood, Shropshire, by Warington Smyth, 1847. On the right are women picking nodules of ironstone from the pit bank. (© Ironbridge Gorge Museum Trust)

44 Calcining kilns of the type developed by the Swedish engineer, John Gjers, who, like many leading engineers of the 1860s, was drawn to Cleveland where the most significant advances in furnace technology were being made. Built similar to a blast furnace, the superstructure of iron-clad brick stands on a sub-structure of cast-iron columns.

Ironmaking

bouts of intense activity, and it therefore became an impromptu meeting place. The Coalbrookdale Co. occasionally remonstrated against 'idle and disorderly people' interrupting the work of the fillers. In the Black Country the topography did not usually allow furnaces to be constructed against a bank. Instead materials were taken from ground level up to the top of the furnace by means of an inclined plane or a vertical lift. Netherton blast furnaces, built in 1813, had a steam-powered inclined plane to haul barrows to the top of the blast furnaces 45ft (13.7m) high. In the eighteenth century raw materials were tipped into the furnace from wicker baskets, but these were gradually superseded by wheeled barrows. In theory the raw materials could be weighed before charging, but in practice it appears that the charge was judged by the barrow load. The proportions of the charge were variable, depending upon the quality of the raw materials and whether the furnace was smelting forge or foundry pig iron. At Coalbrookdale in 1785 François and Alexandre de la Rochefoucauld saw a blast-furnace charge of eight baskets of ore to five of coke and three of limestone. In 1796 Joshua Gilpin visited Ketley Ironworks and saw the furnace charged at a ratio of 7:2:1. At Ketley in 1803 Simon Goodrich saw a charge of 13:5:5.

More accurate figures for blast furnace charges are available for the mid-nineteenth century. At Newland in Furness, which was still smelting with charcoal in the early 1860s, one charge amounted to 415lb (188kg) of iron ore and 400lb (181kg) of charcoal. Over a twelve-hour period between furnace tappings, the furnace was charged sixteen times. In 1851 Russell's Hall furnaces, near Dudley in South Staffordshire, were blown with hot blast at 315C. A single charge comprised 1½ tons of iron ore, 9cwt (457kg) of limestone and 18cwt (914kg) of coke and coal. Every 30cwt of iron ore yielded approximately 14½cwt of pig iron, a yield of 48 per cent. At Cwm Celyn, Monmouthshire, built in 1839, a charge was 7¼ tons of iron ore (equivalent to 18 barrow loads), 2cwt (102kg) of limestone and 3¼ tons of coke (equivalent to just over nine barrow loads).

Blast furnace management was the responsibility of the furnace keeper. Other key roles were in overseeing the charging, tending the blast engine or blowing apparatus, and the founder. To retain optimum performance air had to be blown at the correct pressure, of 3–4psi, and where hot blast was employed, at the correct temperature. The need to tap the blast furnace every twelve hours meant that each furnace was worked by two teams on twelve-hour shifts. The necessity of keeping the furnaces in blast meant a seven-day week, during which time night crews saw little daylight while their two-week shift lasted. The change from night to day work was the most gruelling aspect of work at the blast furnaces because it entailed working a 'double turn' of twenty-four hours. By the early nineteenth century most iron companies stopped the blowing engines for eight hours on a Sunday, which allowed a rest day for the workmen and allowed any maintenance work to be carried out. But the furnaces could only be allowed to stand for a short time. If the furnace cooled too much then the lining would crack, resulting in the shut down of the furnace while its interior was re-lined in refractory bricks. As smelting became more efficient in the nineteenth century the furnaces were tapped more often – by the 1860s the furnaces at Ebbw Vale were tapped every four hours.

The furnace keeper's job was to ensure that the furnace produced iron of a specified quality. Successful management of the blast furnace required fine judgements. Among the most common problems was the conglomeration of material in the furnace, known as 'scaffolding'. It had several causes. If there was insufficient fuel the ore and limestone would not be hot enough to melt consistently. If the coke was poor quality it might be too friable to take the weight of materials in the furnace, crumble and impede the current of air from the tuyeres. If the ironstone or limestone nodules were too large they would not melt and fuse correctly. Each situation would require remedial action before the furnace was working to its optimum capacity.

Before the furnace was tapped the blowing engine was stopped and the molten slag, which being lighter collected on top of the molten iron at the base of the furnace, was taken off. If the slag was too fluid the keeper would conclude that more iron ore was needed in the charge, whereas if the slag was too thick, he would conclude that more limestone was needed. The slag was allowed to run out and form a mass on the floor of the casting house, but some ironworks adopted a system whereby the slag flowed directly into trucks, from where it could be wheeled away to the slag tips. The iron was tapped by unblocking a clay plug in the tap hole, at the base of the furnace. The language used to describe the molten pig iron as it solidified, qualities such as 'creamy', are typically abstruse.

Molten iron was run into sand beds to form the ingots known as pigs, which were doused with sand to accelerate the cooling. A later alternative was to run the iron into cast-iron pig moulds. Most pigs were sent to the running-out fire for re-melting, although once pig-boiling had been adopted it was no longer essential. Refineries, or running-out fires, were usually built close to the blast furnaces because they were blown through tuyeres at the same pressure as the blast furnace, usually using the same engine. Refineries were rectangular hearths with tall stacks, with hearths about 1 by 1.5m. Early running-out fires had brick-lined hearths, but in the nineteenth century it was more common to have the sides and base lined with hollow cast iron, through which cooling water was passed. The disadvantage of using cast-iron plates was the difficulty of sealing the hearth from the water beneath and around it. When molten metal seeped through the gaps and came into contact with the water, it caused a violent explosion. In the 1850s two Dowlais workmen were killed in accidents of this kind. Capacity of running-out fires varied between a half and over a ton of pig iron. The metal, plus a small quantity of iron hammer scale, was laid on a bed of coke, over which another layer of coke was heaped. The blast was introduced only when the iron began to melt, and after about two hours the metal was ready to be run into moulds. After cooling, the resulting 'finer's metal' or 'plate' was broken up ready for the puddling furnaces. At Ebbw Vale ironworks, when pig iron for best cable was to be refined, the coke was first salted by being steeped in brine.

The puddling furnace never changed radically from its original form. Minor improvements were made by substituting cast iron bottoms for the original sand beds, and introducing the circulation of water for cooling. Tools also remained the same. The rabble was a long bar with a hook end, used for stirring the charge. The paddle was a shorter bar with a flat, chisel-like end. The puddler had several of these tools, which were cooled in a water bosh beside the furnace. The iron in the furnace was worked through a small aperture in the furnace door. For charging and balling the iron door was lifted vertically by a chain at the end of a lever, usually operated by an assistant. Another chain raised and lowered the damper on the furnace stack in order to control the intensity of the heat inside the bowl. Puddling furnaces were repaired every week and every six months needed to be completely rebuilt.

Puddling was widely acknowledged as the most difficult of the metallurgical crafts. Quality and output of iron depended upon the skill of the puddler as much as the materials charged into the furnace. At the Botfields' Old Park and Stirchley forges in Shropshire in 1832, individual puddlers produced between 18 and 25 tons of puddled iron over a four-weekly reckoning, earning between £6 and £8 9s for themselves and their assistants. John Percy confessed that, 'I know scarcely any metallurgical operation more interesting to watch than puddling'. Once the furnace was ready a heat took up to one and a half hours, although no more than seven heats would be possible in a twelve-hour turn. Shifts changed over at six in the morning and evening. The night shift was more agreeable as it lessened the strain of working in intense heat.

The charge consisted of about 4cwt (203kg) of iron and 1cwt (51kg) of hammer scale. When the iron was subjected to intense heat the puddler and his assistant took turns to stir the iron, making sure that none of it adhered to the furnace bottom. In just under an hour the iron would have started boiling and had a blue flame on its surface. At this point the iron was stirred more vigorously by both puddler and assistant in turns, as the iron began to form into pasty masses, known in the trade as 'coming to nature'. The iron was almost ready to be removed from the furnace but, having been worked as a single mass of iron, it was broken up into six balls ready for raking out individually. At this point the damper was closed and the bowl allowed to fill with smoke, starving the atmosphere of oxygen, through which iron would have been wasted through oxidation. After about seventy-five minutes the first ball of iron was removed, and over the next ten minutes the remaining five balls would be taken out, all of about 80lb (36kg) in weight. Immediate afterwards the tap hole was unplugged and the slag was allowed to run off into small wagons. Iron-rich, puddling slag was retained and either added to the charge of a puddling furnace to act as a flux or, more commonly, was added to the charge of blast furnaces.

Each ball was manipulated with tongs on to a truck and wheeled quickly to the forge hammer. All puddled iron was hammered, or occasionally put between mechanical squeezers, before rolling, the process known as shingling. As an amorphous mass of iron it was not physically possible to pass the iron directly through a mill. The interstices of the metal contained small particles of slag, and the most effective way of removing it was to place the iron under a heavy hammer, where the slag was expelled in great sparks. By hammering, the mass of iron was welded into a strong rectangular bloom, and it was common knowledge that repeated hammering improved the quality of iron just as kneading improves the quality of bread. Helves were of wood and cast iron, steam hammers invariably of cast iron, but with wrought-iron striking heads. Hammer heads weighed between 1½ and 3½ tons. The heavy nature of the work necessitated frequent replacement of the hammer head and anvil.

From the hammer the bloom was passed through the puddle rolls. These comprised the 'roughing-down rolls', through which the iron was passed through grooves of diminishing section, and the 'finishing rolls', through which the iron was passed through another diminishing series of grooves, to form puddled bar with a tolerably smooth surface. The size of the puddled bar depended upon the eventual use of the iron, and the grooves in the rollers varied accordingly. For merchant iron and rails, the bars were up to 18ft (5.5m) long and near 3in square. If they were intended for rods they would be round in section. For subsequent rolling into sheets or plates the puddled bars needed to be broader, about 5in by ¾in in section, and 14ft (4.3m) long.

Puddled bars were first cut up into short lengths and stacked, or 'piled' in a furnace ready to be brought to a welding heat. Although these furnaces were known as balling furnaces, the term is misleading, and in the nineteenth century the term faded, to be replaced by 'mill', 'heating' or 're-heating' furnaces. These were reverberatory furnaces very similar to puddling furnaces.

In general terms, the rolling mills were similar to the puddle mills, with diminishing series of grooves (*see pp.55, 58*). Iron passed through a mill to produce merchant bar was known as *No.2*, or common, iron, but where it was cut, piled, heated and rolled again it was known as *No.3*, or best, iron. Quality was also determined by the quality of the pig iron. Piles were normally placed in the furnace in the same direction, so that the fibres in the iron would all be pointing in the same direction. By placing bars alternately at right angles, known as cross piling, greater strength was derived from the interweaving of the fibres. Boiler plates were made in this way. In the manufacture of rails, piles were carefully constructed and contained bars of different qualities. The interior of the pile would comprise puddled iron, or occasionally small, cleaned scraps of nails and bolts, while the best quality iron would be found at the base. For the top of the rail the iron was usually prepared from ores with a high level of phosphorus. The resulting hard iron was the best type for resisting the wear of iron wheels but, being comparatively brittle, would not have been suitable for the entire depth of rail. Rails were normally passed through two pairs of rolls. In the first, the iron was reduced to rectangular bars of increasing length. Only in the second mill train were the grooves shaped to the I-section of the rail.

All forms of rolling were labour intensive until strip mills were introduced in the late nineteenth century. The rollerman had to pass the iron through the mill, for which considerable skill was required to prevent the iron becoming skewed. Once it had been passed through the rolls, the catcher had to pass it back over the top of the rolls – known as a dead pass – in order to make another pass through the next groove, and so on (*see p.55*). When the puddled bar was rolled to its ultimate length it was laid on the ground and beaten with wooden mallets in order to keep it straight as it cooled.

Most mill trains were steam powered by the nineteenth century, but not exclusively so. At Cyfarthfa, water power remained in use in the mills for as long as wrought-iron was produced. Rolling-mill plant and its gearing were unprotected machinery that claimed lives and limbs of hapless boys and men. They also required repair and replacement, for which they were hoisted on large cranes. The rolling-mill cylinders and stands were constructed of cast iron. Gradually blemishes appeared on the grooves and periodically the rolls needed to be reworked at a turning mill.

New markets for wrought iron

The new market that emerged for iron rails from the 1820s, and subsequently iron for locomotives and rolling stock, had a major impact on the industry. The earliest contracts for the supply of wrought-iron rails had been for the Stockton & Darlington Railway in 1821. Most of its rails were rolled by small forges in north-east England, such as the Bedlington Ironworks near Newcastle-upon-Tyne, which refined pig iron purchased on the open market. The same firm won orders for rails for the Liverpool & Manchester Railway in 1829, as did the Penydarren Ironworks in Merthyr Tydfil and Bradley & Co. of Stourbridge.

The trade expanded rapidly in the railway boom years of the 1830s and 1840s. By the latter decade the principal rail manufacturing region was South Wales, where Dowlais,

Ironmaking

Cyfarthfa, Plymouth and Rhymney in Glamorgan, Ebbw Vale, Tredegar, Nantyglo, Blaina, Abersychan, Pontnewydd and Blaenavon in Monmouthshire were the leading makers and served a national market. The Rhymney Co., for example, had supplied rails to the London & Birmingham, Hull & Selby, York & North Midland, Glasgow & Ayrshire, and Edinburgh & Glasgow Railways. Dowlais, the largest producer, included the Great Western, Bristol & Exeter and Sheffield & Rotherham Railways among its clients. Midland forges also enjoyed a large share of the regional market for rails. The London & Birmingham Railway had a predominance of suppliers from the Midlands, including Bradley & Co.'s Capponfield and Chillington Ironworks near Wolverhampton, the Tividale Ironworks at Dudley and the Butterley Co. of Alfreton, Derbyshire. Cleveland and north-east England retained a strong rail manufacturing tradition that grew with the expansion of the region's iron industry after 1850. The chief makers included the ironworks of Bolckow, Vaughan & Co., and the Consett Iron Co. Rails would become one of the most important products of the Consett Iron Co., where there were ninety-nine puddling furnaces in 1863 and 139 in 1871.

This trade marked a significant shift in business practice. Finished iron was now supplied direct to customers, as opposed to semi-finished iron that was distributed via merchants, although some merchant houses like Bailey Brothers of Liverpool dealt regularly in rails. It was in the export of rails that merchants played a key role, the chief foreign markets being France, Germany and the USA. South Wales works were the chief suppliers of rails for export until domestic competition reached the required capacity to receive large orders. Dowlais, for example, supplied railways in Germany, Russia, Canada and the USA in the 1830s and 1840s, while both Dowlais and Cyfarthfa supplied rails to the East India Co.

Rails were a high-volume product – by 1836 Dowlais already produced 20,000 tons of rails per annum – and individual contracts were consequently on a large scale. In 1839 the Pentwyn & Golynos Iron Co. won orders for 1,500 tons from the Taff Vale Railway in Wales, and 14,000 tons from Russia, which has been estimated to have occupied the works for over a year. A distinct advantage for rail manufacturers was that rails were rolled to a standard section, thereby simplifying the layout and operation of the rolling mill. However, entry into the rail-making trade required investment in new mills on flat sites. At Dowlais the Big Mill was built in 1830, supplemented by the Little Mill in 1840, both with their own puddling furnaces and steam-powered mill trains, specifically for the market in rails. Rail mills required more powerful engines than had hitherto been required for rolling bar. In 1857–59 the Goat Mill was built at Dowlais, which the works' engineer William Menelaus claimed had three times the power of any existing mill, although its source of power was still a beam engine powered by two high-pressure cylinders. Rail mills also required larger sites because more powerful machinery was intended to facilitate the rolling of progressively greater lengths of rail. For the Great Exhibition of 1851 the Cwmavon Works at Port Talbot rolled a rail 62ft (18.9m) long, while a decade later the new Goat Mill at Dowlais rolled a 120ft (36.6m) rail for the 1862 International Exhibition.

The growth of railways also created a demand for various grades and types of wrought iron as component parts of locomotives and rolling stock. Locomotive manufacture was concentrated in Glasgow, Lancashire and the West Riding, with a scatter of other manufacturing centres, including London, Swindon, Ashford and Bristol. Components such as springs and couplings were made in the West Bromwich area from the 1830s. From a local perspective it characterises a shift towards sophisticated engineering and away from the more technically primitive nail trade that had previously sustained the local economy. There were large Midland ironworks such as Round Oak near Brierley Hill, where locomotives and machinery were manufactured, and the Patent Shaft & Axletree Co. of Old Park, Wednesbury in Staffordshire, who manufactured parts for the railway industry.

The engineers Isambard Kingdom Brunel (1806–1859) in Bristol, John Laird (1805–1874) in Birkenhead, William Fairbairn (1789–1874) in Manchester and London and Robert Napier (1791–1876) in Glasgow all began constructing iron-hulled ships in the 1830s and 1840s. The manufacture of ships' plates became a major industry in the second half of the nineteenth century and was most closely associated with the forges of Scotland. Ships required high volumes of plates and angle iron – Brunel's SS *Great Britain* was built with 800 tons of plates – but the growth of the Scottish forges that supplied them was steady rather than spectacular. A typical early example was the Mossend Iron Co.'s forge at Holytown near Coatbridge, which was built with only eight puddling furnaces in 1840–41 and grew slowly, specialising in plates and angle iron. Until the 1870s the largest Scots puddling forges were the Motherwell and St Rollox forges of the Glasgow Iron Co., with sixty puddling furnaces in 1873. The Blochairn Ironworks at St Rollox near Glasgow, owned by Hanney & Sons from 1867, had large banks of puddling furnaces, fifty in 1873, on either side of the Monkland Canal to exploit the market for ship plates. Its capacity was said to be 1,500 tons of plates per week by the 1870s.

Other Scots forgemasters erected forges to manufacture marine fittings like crank and propeller shafts, stern and rudder posts, in addition to ship and boiler plates. Parkhead Forge in Glasgow, for example, was acquired in 1848 for this purpose by Robert Napier & Sons. The rise of engineering firms in Glasgow and the west of Scotland making bridges, gasworks plant and gasometers, floating docks, landing and promenade piers, allowed a number of other small works to become established in the vicinity, some making as little as 250–300 tons of wrought iron a month. Examples are Merryston (1851), Coats (1854), Rochsolloch (1858), Drumpeller (1859), Phoenix and Clifton (both 1861). In 1873 Scotland was said to have 565 puddling furnaces at twenty-one forges. These forges were generally not integrated with blast furnaces, but were sited with convenient access to the Clyde shipyards. In 1872 only three Scots firms – Baird & Co. at Muirkirk in Ayrshire, William Dixon's Govan Ironworks and the Glasgow Iron Co.'s Motherwell and St Rollox works in Glasgow – operated blast furnaces as well as forges.

Ironworks in north-east England were large-scale concerns, usually integrated with blast furnaces. Rails, ship plates and other marine and engineering work were their principal output. A related trade was the provision of armour plates for naval vessels. By the early 1850s Samuel Beale & Co. of Park Gate Works near Rotherham rolled 4in thick plates for the early ironclads. John Brown & Co. of Atlas Forge, Sheffield, invested £250,000 in 1863 in a new armour-plate mill, capable of rolling armour 12in (0.3m) thick. The previous year it had exhibited armour plates at the International Exhibition, including a single plate weighing over 10½ tons, over 21ft long, 4ft wide, and 6½in thick (6.4 by 1.2 by 0.2m). Armour plates were also manufactured under the hammer by firms such as the Mersey Iron & Steel Co. and the Thames Iron Co. Other South Yorkshire firms such as Low Moor and Bowling Ironworks near Bradford and Farnley Ironworks near Leeds specialised in heavy engineering components, including boiler plates and locomotive tyres. Kirkstall Forge and Wortley Top Forge, the latter a seventeenth-century forge by the River Don near Sheffield, specialised in railway axles (*see p.133*).

Secondary industries

Sheffield dominated the British steel industry from the mid-eighteenth century, but iron and steel were not mutually exclusive. The town owed its position to the technology of crucible steel, invented by Benjamin Huntsman (1704–1776). Huntsman was a clock-maker whose steel springs, manufactured from blister steel, failed because of the inconsistency of the metal, where carbon had diffused very slowly into the surface of the metal. In about 1742 he found a solution – to melt blister steel in a crucible at high temperature, allowing the carbon to disperse more evenly through the metal. He also found that steel could be produced by melting scrap iron and then adding carbon. Steel was produced in melting holes in which crucibles were lowered over a large coke-fuelled furnace chamber, with a tall flue at the rear. The sealed crucibles were pre heated and melted the charge in a process lasting up to five hours at 1,500–1,600C. The metal was then poured into an ingot mould, in which form it was sent for rolling or hammering to produce bar, or drawing into wire. The technology relied upon individual experience and judgement, ensuring that its dissemination was limited. Sheffield became the most important producer of steel cutlery, exploiting a large export market in the USA, and of special steels for machine tools.

The technology inhibited the scale of the industry. Converting furnaces produced steel bars by cementation. But the crucibles, placed in melting holes in the floor of the building, had a limited capacity as they had to be lifted by one workman. In the eighteenth century individual charges slowly increased from Huntsman's 13lb to reach 25lb (5.9–11.3kg), and by the mid-nineteenth century a large charge weighed up to 70lb (31.8kg). This technology allowed small-scale producers to co-exist with larger concerns. In the mid-1850s the Brightside Works was the largest in Sheffield, with ten converting furnaces and 120 melting holes. It was soon to be overtaken by Cammell's Cyclops Works, established in 1846, and John Brown's Atlas Works, established in 1855. Both, like the Park Gate Works, manufactured wrought iron as well as steel.

Another related industry that deserves special mention is the tinplate industry. Tinplate had been manufactured in Britain since the seventeenth century. It was used to make dairy utensils and could be japanned to produce luxury items like tea trays, mugs, inkstands and candlesticks. In Britain the tinplate industry did not exist on any significant scale until the early eighteenth century, when John Hanbury coated rolled iron in tin at his ironworks in Pontypool. In the early nineteenth century there were several short-lived tinplate works, both in the Midlands and South Wales. Hitherto it had been a subsidiary of the iron industry but, when it rapidly expanded from the second quarter of the nineteenth century, largely due to the demands of the packaging industry, it became an industry in its own right. This period of tinplate manufacture was dominated by works in South Wales, concentrated on the west side of the South Wales coalfield, in the Neath and Swansea valleys and further west into Carmarthenshire. Only a minority of these works, notably Ystalyfera at the upper end of the Swansea valley, integrated iron smelting, refining and tinplate manufacture.

Tinplate manufacture operated independently of the major forges, but represented a significant market for pig-iron producers. Vernon Tinplate Works, built in 1850 near Neath, is typical in that it operated its own puddling forge and charcoal fineries. The pattern was a common one in the mid-nineteenth century. 'Coke' and 'charcoal' were the two main grades of tinplate, even in the late nineteenth century when 'charcoal' was merely a brand made from open-hearth steel. Treforest Tinplate Works, near Pontypridd, followed the current of tinplate manufacture in nineteenth-century South Wales. The site had been purchased by the Crawshay family of Cyfarthfa Ironworks because it was well placed to exploit water from the River Taf. William Crawshay II (1788–1867) completed the tinplate works in 1835 and installed his son Francis (1811–1878) as manager. In 1842 it had its own puddling furnaces, as well as four charcoal fineries and two furnaces for melting tin. The works was greatly expanded in 1854, a decade of significant growth in the tinplate industry, when a new tinning house was built with twelve tin furnaces. Difficulties set in towards the end of the nineteenth century, as the technology employed in British works began to become outdated, and the USA imposed tariffs on imported tinplate.

Production of tinplate was a lengthy process of rolling sheets of iron ready for their thin coating of tin, which was practically the last stage in its manufacture. It began with iron bars which were each rolled out into a sheet nearly 2m long, then doubled over and rolled again, and so on until it formed a pack of eight folded sheets. This stage of the procedure was known as hot rolling, and the sheets were known as black plates.

The packs, broken into separate plates, were then ready for cleaning. Until 1843 this was done by scaling, beating the sheets to remove the scale, but was replaced by pickling, which involved submerging the plates in a vat of hot sulphuric acid. Passing the iron repeatedly through the rolls made the metal brittle so its malleability had to be restored by annealing. In their cold state the plates were passed through another set of rolling mills which were designed to give the plates a good smooth surface and even thickness. This cold rolling again made the iron brittle so another annealing was necessary. From the second annealing furnace the plates were once more cleaned in a vat of acid before tinning. The tinning process involved passing the plates through a vat of oil, which acted as a flux, then immersing the sheets in tin, after which they were sorted, cleaned and boxed.

The Midland wrought-iron industry

The Midland forge trade had helped to pioneer the tinplate industry, but it was only a relatively small producer in the nineteenth century, one strand of its particularly diverse iron industry. The Midland region did not compete in the marine engineering market and had only a small share in the rail market. An examination of its character belies the assumption that Britain had a universal national iron industry by the mid-nineteenth century. Ship plates and angle iron were high-volume products and so the most likely source of competition faced by northern and Scots forges came from firms specialising in rail manufacture. The Rhymney Iron Co., for example, was soliciting orders for ship plate and angle iron in 1857–58 in an attempt to diversify its product range away from rails.

A survey of the South Staffordshire iron industry in 1865 recorded ninety-nine forges, of which only twenty had above thirty puddling furnaces. Output was concentrated upon iron of superior quality, including bar, rods, hoops, tinplate and boiler plates, which accounts for the survival of a large number of comparatively small units. Old-established forges such as Wilden near Stourport, which was owned by J.P. and W. Baldwin by 1873, had puddling furnaces, but its principal output remained charcoal iron, mainly

Ironmaking

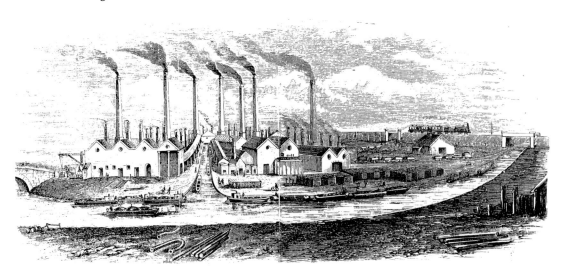

45 Bloomfield Ironworks, Tipton in the Black Country. The sheds house puddling furnaces, as represented by the shorter stacks, while the larger stacks correspond to the engines driving the mills.

for tinplate. Other small forges in the Stour Valley, like Cookley and Broadwaters, also produced iron for the tinplate trade, while Hyde forge manufactured wire rods, plates and bars.

The importance of the charcoal iron trade in the nineteenth century has been overlooked. John Knight (1765–1850) manufactured tinplate and wire rods from charcoal iron from the second decade of the nineteenth century, recognising that his small Wolverley and Cookley forges needed to specialise. Elsewhere in the Midlands, charcoal ironmaking was a component of larger coalfield ironworking businesses. Eardington and Hampton Loade, two forges built in the late eighteenth century close to the River Severn in Shropshire, were acquired in different partnerships by the Stourbridge ironmaster James Foster (1786–1853), in 1809 and 1820 respectively. Both had been stamping and potting forges, but Eardington became a charcoal forge from 1813, as did Hampton Loade from 1826, constituting the charcoal branch of a substantial business empire encompassing smelting, refining and engineering. Activity at Hampton Loade is well documented. It acquired charcoal from at least fourteen different sources, mainly in the Severn and Wye valleys, ranging from Buildwas near Coalbrookdale in the north to Abbey Tintern, 57 miles to the south. It suggests that the decline of charcoal-using rural forges allowed more charcoal to be purchased by fewer works. Between 1829 and 1844, 26,972 tons of charcoal iron were produced at Hampton Loade, an average yearly output of 1,686 tons. The best year was 1835, when 2,721 tons of iron were produced, which should be compared with 1749, when the total output of Shropshire forges was only 2,260 tons.

In the 1820s the Botfields, whose Old Park Ironworks was one of the chief producers of pig and wrought iron in Shropshire, expanded their interests by building new blast furnaces and a forge at Stirchley, and in the following decade by building more blast furnaces at Dark Lane. Investment was also made in refining wrought iron with charcoal. Production began in 1827 and by 1832 there were four charcoal hearths. In 1833, the only year for which there are output figures, 983 tons of charcoal wrought iron were produced at Old Park, which amounted to over 11 per cent of the combined output of the Old Park and Stirchley forges.

Although the number of charcoal-using forges declined, and output declined in relative terms, in real terms it increased in the nineteenth century. Production was focussed upon a smaller number of producers, centred on the coalfields. Other Midland coalfield forges engaged in charcoal iron manufacture included Snedshill in Shropshire, which had forty puddling furnaces and eight charcoal hearths in 1873. Wilden, already mentioned, Darlaston Iron & Steel Works and Monway Ironworks at Wednesbury, both in Staffordshire, were other significant producers of charcoal iron into the 1870s. Iron refined with charcoal was of sufficient importance for large ironmaking concerns to invest in its development. It added to the range of products offered by iron makers, and it highlighted the reputation for quality with which the Midland iron trade sought to identify itself.

Charcoal iron was particularly associated with the wire industry. The market for wire expanded rapidly in the nineteenth century, centred upon Birmingham, with other foci of the industry in Halifax and Warrington. The iron was sold in the form of rods for further drawing, or in coils ready for cutting up for manufacturing small wares. Wire, whether it was drawn from charcoal-refined or puddled iron, was used as the basis for manufacturing a number of products in addition to wire fencing. The most important were telegraph wire, wood screws, cables for suspension bridges and ships' rigging, and smaller wares like springs, bolts, rivets, hooks and eyes, cotter pins, mousetraps, bottling wire, button hooks, knitting needles and skewers. It could be galvanised, tinned or coppered, giving it additional versatility. It was estimated in 1866 that more than 1,000 tons of wire annually were used for making springs for mattresses and other furniture, while 100,000 miles of wire were applied to tying down the corks of soda-water bottles.

Given the Midlands' concentration upon high-quality wrought iron and its limited presence in the major new markets for wrought iron, it is worth questioning whether the success of the iron industry in the late eighteenth and early nineteenth centuries, and the mature and secure market that ensued, ended in entrepreneurial inertia. The close ties that had been developed with customers based on specific products at first seems desirable but might have been an impediment to radical transformation. However, most of the new markets required high-volume output, which probably inhibited existing moderately sized works. The new engineering sections required larger mills with more powerful machinery and served by larger banks of puddling furnaces than were required for merchant bar. Midland forges therefore became characterised by their comparatively small scale, adding to the diversity and complexity of the iron industry. Even as late as the 1870s the British iron industry retained a regional character influenced by location, markets and historical context.

7

CAST IRON AND ENGINEERING

In the nineteenth century cast iron became an industry in its own right. It had always been a specialised branch of ironworking, partly because the foundry was a manufactory in its own right, unlike the forge which supplied bar iron to be manufactured into finished products elsewhere. The arms industry, which has already been referred to, also contrived to create a separate market for the foundry. But in the nineteenth century the crucial factor in the development of cast iron was technology.

Previous to the eighteenth century, iron was cast direct from the blast furnace. From the early eighteenth century the use of air furnaces had allowed the foundry to work independently of smelting. Of specialist eighteenth-century foundries relying upon air furnaces, Bersham, near Wrexham in North Wales, is a good example. Its lease was acquired by Isaac Wilkinson in 1753, inherited by his son John in 1763, who was later in partnership with his brother William. After John Wilkinson acquired his patent for boring cannon in 1774 the works concentrated on producing cannon and engine cylinders and a new foundry was built. When Marchant de la Houliere visited Bersham in 1775 he noted that the foundry had four air furnaces which cast iron simultaneously into a single central pit. The comparative small capacity of the individual furnaces necessitated the building of an octagonal foundry, in which furnaces ranged around the sides could be operated in unison for the casting of large objects.

Some of the more illustrious products of eighteenth-century foundries – cannon, bridges, engines – have already been described. A wide range of other products was also cast by foundries in the eighteenth century. The Carron Co. cast stoves and grates, agricultural ironwork like plough plates, as well as pots and pans. By the late eighteenth century it was exploiting the market for sugar pans, shipped to the West Indies via Glasgow. Other types of machinery included printing machinery, weighing machines and cotton manufacturing equipment. John Smeaton built a boring mill there and the company hired Sussex men to develop Carron's ordnance trade, which was the most important branch of its business to 1815. In 1796 the output of the foundry at Bedlam, Shropshire, included anvils, frameworks for mine pulleys, crank shafts and engine pistons, plates for furnaces, tram wheels and iron rollers for an inclined plane. Among the final items cast before the foundry closed seven years later were parts for engines, tramroad plates and the frame of a waterwheel.

To expand, the foundry and engineering industries needed a furnace of larger capacity than the air furnace, which it found with the invention of the cupola furnace, patented by William Wilkinson (1744–1808) in 1793. A cupola is a simple brick shaft furnace charged at the top and tapped at the bottom. Early examples were small, only 10ft (3m) tall, and strengthened with wrought iron hoops. One such cupola was observed

Cast iron and engineering

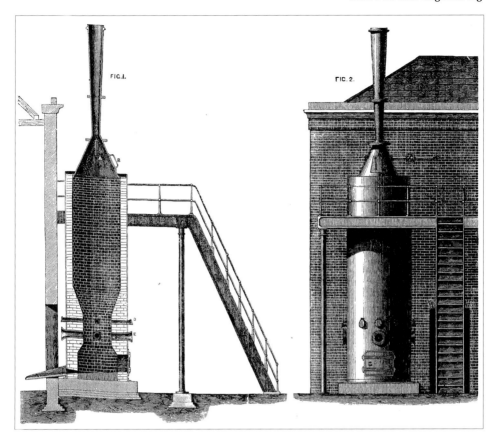

46 Elevation and cross-section of a cupola furnace, showing its tall slender proportions, tuyeres to convey the blast, charging level at the top and taphole at the base.

in Hull by the Swedish industrialist Eric Svedenstierna in 1803 where it was blown by horse-powered bellows. Later cupolas were taller, and clad in wrought iron or steel sheets, thinner but otherwise similar in appearance to contemporary blast furnaces. Cupolas have remained the standard form of melting iron for the foundry and as such they have survived into the twenty-first century with only minor alterations, such as electrical-powered hoists and blowing machines and the control of emissions by closing the tops.

Moulds for small items were made up in wooden boxes, or flasks. Patterns of wood or, later, steel were laid out on a wooden board. On to this was laid a wooden frame forming one half of the box, which was filled with sand and compacted with a mallet. Then the whole was turned over, the wooden board removed, and another frame was laid on and filled with sand, enclosing the patterns in the middle. Finally the two halves of the framework were removed in order to lay the patterns back on to the board and the two halves of the box were reunited. Larger items could be cast into sand moulds on the floor of the foundry. In the early nineteenth century items such as engine cylinders were still cast by loam moulding. Loam was a mixture of sand, clay, straw and horse manure, or some other binder pasted over a former to the approximate shape required, then smoothed off and dried.

Ironmaking

47 Making a casting box. Patterns are laid out on a board ready to be pressed into a box of sand. (© Ironbridge Gorge Museum Trust)

Unlike a blast furnace, a cupola was not kept continuously in blast. Fuelled with coke, charged with pig iron or scrap, sometimes using limestone as a flux, and blown from a waterwheel or steam engine through tuyeres, its operation was nevertheless similar. The description written by Joshua Gilpin at Coalbrookdale in 1796 of the tapping of a blast furnace for casting is equally applicable to the tapping of a cupola furnace, at least in the casting of small items:

> When ye workmen hear the bell each moulder from the various rooms runs with a ladle of iron to catch a due quantity of metal & takes it to his mould which he fills with it – a mould is always to be filled at one pouring; if one man's ladle will not hold enough at a time a large one is brought carried by two or more persons & if a very large casting, the metal is run into a large pot & both hoisted & carried to the mould by a crane.

In such cases, tapping could continue for another hour.

Cupola furnaces allowed much larger items to be cast, and consequently new forms were made possible, especially in the field of structural sections. Cast iron was favoured for bridges for as long as the arch was the primary structural form. Coalbrookdale set the pace in the early years. It cast a bridge for Capability Brown's landscape park at Trentham, Staffordshire, in 1794, and in 1795 ribs were cast for a new bridge at Buildwas, spanning the River Severn just upstream of Coalbrookdale, designed by Thomas Telford (1757–1834). Whereas the original Iron Bridge had used 378 tons of cast iron, Buildwas bridge used only 170 tons in a 130ft (40m) span. Many of the surviving early iron bridges are in South Wales, where they were built for tramroads conveying raw materials and finished iron, and they belie the image of South Wales as just a region of mass-produced

Cast iron and engineering

48 *Above* Tapping a cupola furnace by filling ladles placed beneath a runner from the tap hole. Tapping would continue for up to an hour, during which time each founder filled his ladle many times.
(© Ironbridge Gorge Museum Trust)

49 *Right* Filling casting boxes in a foundry. Between six and eight boxes could be filled from a single ladle.
(© Ironbridge Gorge Museum Trust)

Ironmaking

50 Pontcysyllte Aqueduct, by William Jessop and Thomas Telford, carried the Ellesmere Canal over the River Dee by means of iron arches and an iron trough on stone piers. It opened in 1805 and was one of the most ambitious feats of civil engineering in iron. (© author)

51 A nineteenth-century cast-iron bridge on the Dudley Canal in the Black Country. (© author)

puddled iron. By the early nineteenth century several firms, such as the Butterley Co. of Derbyshire and Horseley Ironworks in Staffordshire, specialised in bridge castings for roads, canals and early railways. Engineers quickly gained confidence in the use of iron for aqueducts, as is demonstrated by the spectacular Pontcysyllte Aqueduct, on the Ellesmere Canal in Denbighshire, a cast-iron trough 1,000ft (308m) long designed by William Jessop (1745–1814) and Thomas Telford and completed in 1805.

The Butterley Co. successfully exploited a growing market for structural ironwork, supplying the cast-iron components of bridges, lock gates and cranes at London, Bristol and other docks. Butterley provided cast-iron roof trusses for warehouses, including in

1817 trusses of 54ft (16.5m) span for a mahogany warehouse at West India Dock, London. The company also provided cast-iron rails for early railways, including at London docks, and water pipes, demand for which grew in the early nineteenth century for waterworks and gasworks. It won contracts partly because it developed close relationships with engineers such as John Rennie (1761–1821), who had a high regard for Butterley iron, and William Jessop, who was a partner in the iron company. It also employed skilled engineers. William Brunton (1777–1851) was engineer from 1808 to 1821 and pioneered marine and locomotive engine manufacture at Butterley. His expertise was entirely empirical, having learned mechanics from his father and grandfather, followed by stints working for foundries at New Lanark and Soho, Birmingham. His successor, Joseph Glynn, was more scientific, and eventually became a Fellow of the Royal Society and Member of the Institution of Mechanical Engineers.

Butterley cast and assembled a variety of other iron goods, including engines. It was one of a number of ironworks to take an interest in engine manufacture following the expiry of James Watt's patent in 1800. New forms of beam engine producing steam at high pressure were pioneered, notably by the Cornish engineers Arthur Woolf (c.1766–1837) and Richard Trevithick (1771–1833). Trevithick had been associated with numerous ironmasters in north-east England, in the East Shropshire Coalfield, and at Merthyr Tydfil. Coalbrookdale and Penydarren manufactured high-pressure engines, and at the latter works a steam locomotive was manufactured that was the first locomotive to run on rails when, in 1804, it completed a trial journey on the Penydarren Tramroad south of Merthyr.

Trevithick also had engines made at Neath Abbey in the period 1804–06. Blast furnaces had been built at Neath Abbey in 1793 and soon after 1800 it began its specialisation in engine manufacture, in which it gained a widespread reputation under the direction of Joseph Tregelles Price (1786–1854). At first, its principal market was the Cornish mining industry but from the 1820s it served the more lucrative market for engines in the coal, iron and tinplate industries of South Wales. The business expanded to the extent that in 1845 the Neath Abbey Iron Co. built new blast furnaces further up the Neath valley at Abernant to supply pig iron to the foundry and engineering works, the Neath Abbey furnaces having ceased work. In the 1850s Neath Abbey built 119 engines, of which 102 were stationary steam engines, five were locomotives and ten were marine engines. Of the stationary engines only fourteen were beam engines, the remainder being horizontal or vertical-mounted engines. But the heyday was relatively short-lived. Smelting at Abernant ceased in 1861 and, although the company went into a slow decline, the foundry and engineering side of the business outlived the smelting sector upon which it had been established. The pattern was repeated elsewhere.

Widespread adoption of the cupola allowed the separation of smelting from foundry work. In the long term it allowed foundries to be set up independent of smelting. Boulton & Watt's Soho foundry was built in 1796, where previously they had been engineers contracting the foundry work required for their engines to already established iron companies. Murray, Fenton & Wood's Round Foundry in Leeds was built in 1795. It made flax machinery and iron frames for flax mills, then steam engines and locomotives from 1812. Engine manufacture became established at foundries in Cornwall, like Perran Foundry and Harvey & Co. of Hayle, who could meet the demand of the region's mining industry. Then they branched out, making marine engines and railway

Ironmaking

locomotives by the third decade of the nineteenth century. Henry Maudslay (1771–1831) set up in business in 1797, and quickly gained a reputation for manufacturing machine tools. He established a successful business in south London, later joined by his sons and Joshua Field (1786–1863), who traded as Maudslay Sons & Field after 1831 and built a reputation for manufacturing machine tools. William Marshall's Britannia Ironworks in Gainsborough cast plant for cake mills and linseed oil mills, but made its mark in the agricultural sector, manufacturing portable engines, threshing machinery, saw benches, pumps and straw elevators, for which there was a rapidly growing export market in the nineteenth century.

Advances in wrought-iron structural sections such as the girder, and in suspension bridge engineering, saw a relative decline in the proportion of cast-iron bridges. The engineering industry required both cast and wrought iron. William Hazledine (1763–1840) had foundries at Shrewsbury and Plas Kynaston near Wrexham, and a forge at Upton in Shropshire. He provided the cast iron for Pontcysyllte Aqueduct and later provided the iron, including wrought-iron chains, for Telford's Menai suspension bridge, opened in 1826. Shipbuilders such as Robert Napier & Sons in Glasgow had foundries and forges including Vulcan foundry and Parkhead Forge.

Cast iron was an ideal pre-fabricated material. Foundries could produce relatively cheap architectural components like classical style columns and traceried windows by

52 Advert of *c*.1860 for portable engines and threshing machines manufactured by William Marshall's Britannia Ironworks in Gainsborough.

the end of the eighteenth century, and in the nineteenth century made ironwork for roofs, including glass houses, which gave a further fillip to the structural and aesthetic possibilities of iron. Architects took to it quickly. John Nash (1752–1835) used cast iron structurally and decoratively for the Brighton Pavilion (1815) and used it profusely for railings, gates, lamp holders and balconies in town houses, as did Robert Adam (1728–1792). The proliferation of urban parks in Victorian Britain created a demand for bandstands, ornamental fountains and gates. Not much of this work, and thousands of tons of it was cast in the nineteenth century, was at the cutting edge of artistic taste. Cast-iron gates replicated the Baroque style of eighteenth-century wrought-iron gates, although much heavier in appearance. Fountains drew on a wide repertoire of classical forms, and even mythology. Some firms specialised in it. The Coalbrookdale Co. specialised in art castings when the ironworks were no longer able to compete in the smelting or forging sectors, and established a high reputation for its work, not least on the back of its success at the 1851 Great Exhibition.

But most of the nineteenth-century art castings came from Scottish foundries, from where the ores were particularly suitable for producing finely moulded work. Such foundries as Carron Co. and McDowall Stevens & Co. of Falkirk, Macfarlane & Co. of Saracen Foundry in Glasgow, as well as Andrew Handyside of Derby, prospered into the twentieth century by casting some of the signature motifs of British life – cast-iron pillar boxes and telephone call boxes like the K6 designed by Giles Gilbert Scott (1880–1960) and introduced in 1936.

53 A park fountain commemorating Queen Victoria, erected in Vivary Park, Taunton, in 1907. It was cast by Macfarlane & Co. of Saracen Foundry, Glasgow, well-known makers of municipal park furniture. (© author)

8

IRONMASTERS

Ironmasters and managers

The capitalist structure of the iron industry emerged in the sixteenth century. From it came the ironmasters who were to enjoy increasing power and wealth when the industry expanded at the end of the eighteenth century. Ironmasters, not inventors, were the individuals who dominated the course of the iron industry well into the nineteenth century, until they were eventually overtaken by limited companies formed under the Companies Act of 1862.

From the mid-sixteenth century the Crown disposed of estates previously held by the monasteries, and new landowners pioneered their industrial exploitation. Although landowners were associated with many of the earlier blast furnaces outside of the Weald, such as the Forest of Dean ironworks built by the Earl of Pembroke in the early seventeenth century, which have already been described, owners soon withdrew from direct involvement in the industry. Instead, they leased land and mineral reserves to professional ironmasters.

Ironmaster is an imprecise term. The high capital outlay necessary to construct a works encouraged the formation of partnerships. In theory, any of the investors could describe themselves as an ironmaster but in any partnership there were usually only one or two men actively engaged in running the business. When the two Bedlam blast furnaces were built in Shropshire in 1757, the capital was provided by a partnership of twelve equal shares owned by nine men. They included the landowner, John Smitheman, while other members had interests in the coal trade or simply saw the partnership as a way of investing capital. Only one partner, Edmund Ford, who was also a partner in nearby Leighton blast furnace, described himself as an ironmaster.

As late as the eighteenth century ironmasters were not exclusively interested in iron, as their enterprises also encompassed non-ferrous metals, of which the examples of Abraham Darby I and William Wood have already been described. In the nineteenth century ironmasters would develop other subsidiary industries such as coal mining and brick and tile manufacture which, in many cases, eventually superseded iron in importance. Crawshay Bailey (1789–1872) started his career working for his uncle Richard Crawshay at Cyfarthfa, and in around 1820 joined his brother as a partner in the Nantyglo Ironworks. Profits from ironmaking were invested in agricultural estates in South Wales, including in the Rhondda and Cynon valleys. From the 1840s, coal from these estates became one of the principal sources of his personal wealth, until 1867 when he sold out for a handsome profit to the Powell Duffryn Steam Co.

An ironmaster generally took on the responsibility of finding markets and negotiating prices and credit agreements with iron merchants or direct customers. An ironmaster may also have supervised production where an ironworks comprised a single furnace or forge. Ironmasters such as Richard Knight may have started thus at Moreton Corbet in Shropshire, before he acquired too many geographically separated interests. As the ironmasters became responsible for increasing numbers of works, it is self-evident that they would have been unable to supervise day-to-day operations, resulting in the delegation of managerial responsibilities. This created a role for managers, sometimes referred to as the clerk or agent. The manager organised production, expedited its transportation, acted as cashier responsible for collecting payments, and hired and paid workmen.

The position of salaried manager was the channel through which many men, such as John Wheeler of the Ironworks in Partnership in the seventeenth century, rose to the status of partner and became ironmasters in their own right. John Guest (*c*.1721–1787), a founder from Broseley, Shropshire, was appointed works manager at Dowlais in 1767, bringing two of his younger brothers with him. Dowlais had a large number of small shareholders, so Guest was able gradually to buy his way into the partnership and become one of the largest shareholders. His grandson, later Sir Josiah John Guest MP (1785–1852), was at the head of a concern that eventually grew into the largest ironworks in Britain. Other men entered industrial capitalism having acted as landowners' agents. In Shropshire, William Ferriday father and son (1697–1756 and 1737–1801 respectively) were agents to the Forrester estates in the 1750s and became partners in three of the local ironworks; Thomas Botfield was agent to Isaac Hawkins Browne, on whose ground he built the Old Park ironworks in 1790.

Growth of integrated ironworks from the late eighteenth century demanded additional tiers of management. Supervisors or overseers were required for the mines and collieries, the blast furnace, the forges, rolling mills and foundries. Usually one person retained an overall managerial role, and remained the intermediary between ironmaster and workmen. John Gibbons (1777–1851), ironmaster of Level and Corbyns Hall Ironworks in Staffordshire, described the three main attributes required of such a man to be 'character, capacity and technical knowledge'. Neither managers nor ironmasters were themselves capable of performing skilled work at the furnaces, forges and mills, but they had to deal with strong-willed and well-informed men who could.

Like the role of ironmaster, managerial roles created their own dynasties. The Gilpin family entered the coal and iron trades in Cumbria in the seventeenth centuries. Benjamin Gilpin (d.1793) left Cumbria with Isaac Wilkinson when he moved to Bersham in 1753. His son was Gilbert Gilpin (1766–1827), manager first of Bersham, later of Boulton & Watt's Soho foundry, Sirhowy Ironworks and then of the Old Park Ironworks in Shropshire from 1800 to 1814. Another branch of the family left Cumbria for Shropshire: Mark Gilpin (d.1805) was manager of the Coalbrookdale Ironworks for Abraham Darby III, while his son became book keeper at Penydarren.

Ironmasters in society

Ironmasters were part of the new capitalist class that emerged in the eighteenth century. Much has been written about the contrast between trade and the gentry during this period, but the opposition of landowners and businessmen can be misleading and was

Ironmaking

never so clear cut. Landowners did not live in a parallel universe of the country estate and were never slow to cash in where their property contained valuable mineral reserves. During the exponential growth of the industry from the end of the eighteenth century ironmasters enjoyed an innate self-confidence. Ironmasters and their counterparts in other industries saw themselves as the bedrock of national prosperity, the source of Britain's greatness and the means of its defence. Political radicals were hardly to the fore; most ironmasters counted upon the stability of a society ruled by a landed oligarchy in order to safeguard their trade and open up new colonial markets.

Eighteenth-century growth of the iron industry in Britain's rural districts also allied the iron industry to the movement for agricultural improvement and economic development. Socially and politically, ironmasters had much in common with land agents, surveyors and lawyers. Ironmasters even bought farms, giving them land for pasturing horses and exerting control over the local economy to ensure that produce was not sent away to market when the harvest was poor. They also had access to the landed gentry, who were either their landlords, or were the men the industry lobbied in order to have its interests represented in parliament. Edward Kendall (1750–1807), ironmaster of Beaufort, the Monmouthshire ironworks named in honour of its landlord, the Duke of Beaufort, was an experienced and successful lobbyist on behalf of the regional industry, having convinced Members to pass an act for the building of the Monmouthshire Canal in the 1790s. Nevertheless, not every ironmaster would have passed off as a gentleman. Numerous Quaker ironmasters were hardly establishment figures. At the other end of the scale, the aggressive and hard-headed John Wilkinson was unconventional throughout his life, and was outspoken in his support of the French Revolution. A man's man, he fathered three children in his seventies by a servant girl, and had an iron coffin built for himself (which turned out to be too small when the crucial moment came).

54 John Wilkinson (1728–1808), from one of the trade tokens issued in the period 1788–1795. The edges were inscribed with the names of his ironworks – Bersham, Bradley, Snedshill and Willey. (© Ironbridge Gorge Museum Trust)

Ironmasters

The iron industry was a comparatively closed community and the status of ironmaster was passed down through generations. It bred a conservative outlook and one that was difficult for outsiders to break into – even though it was usually outsiders who achieved significant technological advances. A niggling sense of exclusion may have lain behind Richard Crawshay's strident declarations of self-importance. At Cyfarthfa, Crawshay had achieved his ambition to become the nation's leading ironmaster. To celebrate, he commissioned a series of portraits of the 'great iron-founders' of the kingdom, namely himself, his son William, and his friends John Wilkinson and William Reynolds. Placed on the walls of his house in Merthyr, where he received numerous visitors connected with the iron trade, the portraits served a twofold purpose – to boast of his success and to slight his chief rivals, Samuel and Jeremiah Homfray of Penydarren.

The Homfray family were well-established in the Black Country and South Wales, and offer a good case study of how the status of ironmaster was passed down through the generations. The first Francis Homfray (1674–1737) moved from Wales to Swinford, where he worked as a manager for Ambrose Crowley. Later he set up in business in his own right and had interests in two forges and a slitting mill in Staffordshire. After his death his concerns were carried on initially by his wife Mary before passing largely to their eldest son, also Francis (1725–1798). A younger son, John Homfray (1731–1760), established Broadwaters forge near Kidderminster in 1753, which later on passed to his son Jeston (1752–1816), after whose death it was taken by his son George (1778–1848).

The second Francis continued running forges in Staffordshire and in 1782 worked a small cannon foundry at Cyfarthfa, where he briefly leased the stamping-and-potting

55 William Reynolds (1758–1803), attributed to William Hobday, 1796. Reynolds is holding a drawing of the iron aqueduct at nearby Longdon-on-Tern, and in the background is an inclined plane.
(© Ironbridge Gorge Museum Trust)

forge in 1783. His sons were Francis (1752–1809), Jeremiah (1759–1833), Samuel (1762–1822) and Thomas. From 1781 the third Francis Homfray was a partner in a rolling and slitting mill at Hyde, Staffordshire, and in 1781 he acquired the lease of Lightmoor Ironworks in Shropshire in partnership with his cousin John Homfray (1759–1827, brother of Jeston and who later changed his name to Addenbrooke). The best known of the brothers were Samuel and Jeremiah, owners of the Penydarren Ironworks established by their father in 1785, initially with their brother Thomas as another partner. The brothers quarrelled and eventually went their separate ways, with very different consequences. Jeremiah remained an archetypal 'improver', always ready to exploit natural resources in hitherto unpromising territory. He was a founding partner in 1789 of Ebbw Vale Ironworks, and moved into the coal trade in the early nineteenth century. But his luck ran out and he suffered bankruptcy in 1813, following which he lived the remainder of his life in France. Thomas Homfray was also a bankrupt in 1819, having taken Hyde forge from his nephew Jeremiah.

Samuel was the luckiest of the Homfray brothers. In 1793 he married Jane Morgan, daughter of Lord Tredegar, head of one of the principal gentry families in South Wales. In many respects he seems an unlikely candidate for elevation to elite society. Posterity has cast Samuel Homfray as abrasive and arrogant, with an inclination to opulence that crossed the border into vulgarity. According to Gilbert Gilpin:

> His carriage is the most elegant, & daubed all over with armorial bearings, of which he has got a pretty good collection since his marriage into the Tredegar family. He cannot ride into Merthyr without having two livery servants perched up behind, turned up with yellow & silvered just like the doughy kings and queens which we frequently see on a gingerbread stand.

Samuel made further headway into society when he became MP for Stafford in 1818. He no longer had a direct interest in the running of Penydarren Ironworks after 1813. As the last of the great Merthyr ironworks it was comparatively poorly endowed with mineral reserves. Instead he had been able to build a new ironworks on much more favourable terms. Family connections enabled him to obtain a cheap lease of 3,000 acres of ground at the head of the Sirhowy valley, Monmouthshire, for an annual rent of only £500. Here he established the Tredegar Ironworks, which was inherited by his son, Samuel George Homfray (1795–1882), the last member of the family to have been associated with the iron trade. The latter Homfray rebuilt Bedwellty House, a minor residence of the Morgan family, in 1825 as an elegant Regency country house with gardens. It contrasted with the home of his father, Penydarren House in Merthyr Tydfil, an extravagant mansion overlooking the works, and the change symbolised a more general shift in taste and social status.

The Darby and Reynolds families, who controlled the Coalbrookdale group of ironworks for most of the eighteenth century, came from a strong Quaker business community in Bristol. Investment in the Coalbrookdale works was made by fellow Bristol Quakers, notably Thomas Goldney (1664–1731) and his son, also Thomas (1696–1768), prosperous Bristol merchants whose profits from the West Africa and American trade funded the building of Horsehay and Ketley ironworks in the 1750s by Abraham Darby II. Shortly afterwards, Richard Reynolds (1735–1816), another Bristol merchant, devout Quaker and son-in-law of Abraham II, moved to Coalbrookdale. His investment in the local iron industry proved even more lucrative. It was the peak of Quaker influence in the

Ironmasters

Shropshire iron trade, of ironmasters distinguished by their frugality, strict self-discipline, as well as their antiquated dress and mode of speech. The following generations were born into successful business families, and consequently had different outlooks. William Reynolds was the outstanding figure in the Shropshire iron industry by 1800, and also had interests in the chemicals and pottery industries, but he died relatively young (see p.97). His half-brother Joseph left a bitter legacy in the coalfield when he closed Ketley Ironworks following a depression in the iron trade in 1818 and retired to Bristol.

Later members of the Darby family also differed from previous generations. Abraham Darby III (1750–1789) tried to live up to the family name by promoting and building the iron bridge across the River Severn, and rebuilding the Coalbrookdale ironworks, but he was not an astute businessman. The last generation of the Darby family to take a direct interest in the local iron trade were Abraham Darby IV (1804–1878) and his brother Alfred (1807–1852). But the brothers' main business interests were transferred to South Wales when, in 1844, they were leading partners in a new venture that purchased Ebbw Vale and Sirhowy Ironworks, then soon acquired Victoria, Abersychan and Pentwyn, forming a knot of ironworks in upland Monmouthshire. Abraham converted to the Anglican Church and, in 1851, was the principal patron of a parish church built at Coalbrookdale. In the same year he moved to a country house at Stoke Court in Berkshire, while his brother purchased Stanley Hall, a Jacobean country house near Bridgnorth, Shropshire.

It was a familiar pattern. In the eighteenth century, ironmasters could be expected to live in close proximity to their ironworks, but in the nineteenth century their descendants moved out and spent their fortunes on land and the trappings of country

56 Richard Reynolds (1735–1816), who took over management of Coalbrookdale Co. in 1763 and eventually became its landlord. (© Ironbridge Gorge Museum Trust)

Ironmaking

life. In Merthyr Tydfil Richard Crawshay lived in Anthony Bacon's Cyfarthfa House, next to which were built Henry Cort's original puddling forge and the head of the Glamorganshire Canal. His son, William I, continued to live in London after his father died, leaving the management of Cyfarthfa to his son William II. It was William II who built Cyfarthfa Castle in 1825, one of the last but arguably the best example of an ironmaster's mansion overlooking his empire. His father disapproved of this extravagant waste of capital reserves – the house cost £30,000. Robert Lugar, architect of Cyfarthfa Castle, then landed another plum commission, to build a retirement country house for Richard Crawshay's nephew, Joseph Bailey (1783–1858) of Nantyglo Ironworks. Glanusk Park was in the more peaceful surroundings of the Usk valley in Breconshire.

Ironmasters were among the band of new country house builders of Victorian Britain. The Knight family is an early instance of social mobility in the iron industry, in which capital was invested in cultivating a different social environment where the association with trade gradually diminished. Five generations of the Knight family were ironmasters in the Midlands, beginning with Richard Knight (1659–1745). His eldest son, also Richard (1693–1765) purchased Croft Castle, a country house in Herefordshire, in 1746 and his second son, Thomas (1697–1764), was a clergyman. Only the younger sons, Edward (1699–1780) and Ralph (1703–1754), remained in the iron trade and lived close to their respective ironworks at Wolverley and Bringewood. The family's landed estates were inherited in 1765 by Richard Payne Knight (1750–1824), son of Thomas, who had no direct experience in the industry and who enjoyed an uneasy relationship with his ironmaster uncle, and cousins James and John. Payne Knight was a gentleman.

57 Dale House and Rosehill House, built in the second and third decades of the eighteenth century for the Coalbrookdale ironmasters Abraham Darby I and Richard Ford. The lake in front is in fact the reservoir for the Coalbrookdale Upper Works waterwheels. The iron footbridge, built in the 1780s and therefore one of the earliest such bridges, was demolished in 1849. (© Ironbridge Gorge Museum Trust)

Ironmasters

58 *Right* William Crawshay II (1788–1867) of Cyfarthfa Castle, who became head of the family business from 1834.

59 Cyfarthfa Castle, Merthyr Tydfil, by Henry Gastineau, *c.*1830..

Ironmaking

He had completed the Grand Tour, was a connoisseur of landscape who could boast a large number of Claude and Poussin paintings in his collection, and was an influential figure in the Picturesque movement. The first Richard Knight's house at Downton near Bringewood furnace and forge was transformed by his grandson into a country house with landscape garden. Downton Castle, one of the earliest country houses in the castellated style, was designed largely by Payne Knight himself. In his park he created woodland walks along the River Teme where he could appreciate the untamed forces of nature. He kept a head for business too. Estate woodlands turned a healthy profit supplying charcoal to the family forges until, when he was unable to attract a new tenant, Payne Knight closed Bringewood forge in 1815 and incorporated it into the grounds of Downton, turning the reservoirs into lakes with waterfalls.

During the nineteenth century the more sophisticated tastes of the gentry were acquired by the ironmasters, a social phenomenon well represented in the life and works of G.T. Clark (1809–1898). Clark trained as a surgeon but never practised. He began his career as an engineer, working under Brunel on the Great Western Railway, and later worked as an Inspector for the General Board of Health. In 1852 he became trustee of the will of Sir Josiah John Guest, owner of the Dowlais Ironworks and, despite having no previous direct experience in the industry, was in control of the Dowlais Iron Co. until 1897. During Clark's time Dowlais converted to steel manufacture, new sources of ore were exploited and imported from Spain, and a new works was built at East Moors in Cardiff. He also became a prominent public figure. As an active antiquary, Clark was a leading member of the Royal Archaeological Institute and the Cambrian Archaeological Association. His background in engineering stimulated his special interest in military architecture, but he was also a patient compiler of Glamorgan pedigrees, publishing *Limbus Patrum Morganiae et Glamorganiae* in 1886.

Clark purchased Talygarn, a moderate-sized house near Pontyclun in the Vale of Glamorgan, in 1865, and incorporated it into a larger new country house to his own

60 G.T. Clark (1809–1898), scholar-ironmaster and trustee of the Dowlais Ironworks from 1852.

design. Gothic on the outside, Clark's informed Renaissance tastes were reserved for the sumptuous interior. The corridor is enriched by painted ceilings, copies of sixteenth-century Venetian paintings. He had woodwork carved by Biraghi of Venice, and marble chimneypieces by Terrazzi of Verona. Iron gates – shipped from Venice not Dowlais – graced the formal garden. Clark completed his work at Talygarn by building a new church, again to his own design, in 1887.

Masters and men

When they lived close to the works, ironmasters were in effect local patriarchs. They controlled local economies, were often magistrates and Members of Parliament, with a responsibility for poor relief and the welfare of their workmen. From below, ironmasters must have seemed omnipotent, but they were strong characters who could socialise with landowners and with the workmen in a relationship that was not devoid of respect. In spite of his bruising manner, it is said that 20,000 people walked 25 miles from Merthyr Tydfil to Cardiff in 1810 to attend the funeral of Richard Crawshay in Llandaf Cathedral. At his funeral in 1807 Edward Kendall's coffin was carried to his grave by six of his workmen. John Wilkinson made many enemies in his career but remained a hero to many of his workmen. They celebrated his real achievements, and added a few others, in a well-known patriotic song:

>Before I proceed with my lingo
>You shall all drink my toast in a bumper of stingo,
>Fill up and without any further parade,
>John Wilkinson, boys, the supporter of trade.

>May all his endeavours be crowned with success
>And his works ever growing prosperity bless,
>May his comforts increase with the length of his days
>And his fame shine as bright as his furnace's blaze.

>That the wood of old England would fail, did appear,
>And tho' iron was scarce because charcoal was dear,
>By puddling and stamping he cured that evil,
>So the Swedes and the Russians may go to the devil.

>Our thundering cannon, too frequently burst,
>As mischief so great he prevented the first,
>And now it is well known they never miscarry
>But drive on our foes with a blast of Old Harry.

>Then let each jolly fellow take hold of his glass
>And drink to the health of his friend and his lass.
>May we always have plenty of stingo and pence,
>And Wilkinson's flame blaze a thousand years hence.

9

IRON WORKMEN

Just as a new capitalist structure coincided with the dissemination of furnace and forge technology in the sixteenth century, so it was accompanied by a new workforce structure. Although there were similarities between the skills needed in the bloomery and in the finery forge, there is strong evidence that the introduction of the indirect process was achieved by immigrant workmen. Between the 1490s and 1540s over 500 furnace and forge workmen arrived in the Kent and Sussex Weald, mostly from the *Pays de* Bray in Normandy. Their descendants continued to diffuse the new technology throughout Britain, even if the dynamic of iron industry expansion was essentially capitalist.

The principal men at the blast furnace were the furnace keeper and the founder. These roles changed little and were still recognisable in the nineteenth century. The same is broadly the case with the forge, where the principal men were the finer and the hammerman. The finer was responsible for refining the iron in the finery hearth while the hammerman was responsible for drawing it into bars. Both terms had a historical context, having been used to describe workmen at bloomeries. Inventories of finery forges indicate that all of the tools and implements were the property of the works and not the workmen. This was to remain the case in the puddling era.

Before the nineteenth century the number of men employed at a furnace or a forge remained small. To function, a blast furnace needed a mine burner, filler, keeper and founder. A forge with two finery hearths would need two finers, a hammerman, carpenter, smith and clerk. In each case a core group of skilled men could call upon whatever unskilled assistance they needed. In Richard Crawshay's phrase, iron production depended upon 'active and powerful men', a factor that contributed to the industry's macho image. The most skilled men, especially in the forge, needed to be at the peak of their physical as well as mental powers. Boys entered the trade in the hope of rising to the status of finer or hammerman, puddler or rollerman, while men in physical decline withdrew from the limelight to find less strenuous tasks.

Puddlers, and presumably finers before them, began to feel the strain of their workload once they reached forty, although a few men continued to perform their trade into their fifties. According to John Percy:

> the majority [of puddlers] die between the ages of 45 and 50 years; and, according to the returns of the Registrars, pneumonia, or inflammation of the lungs, is the most frequent cause of their death. This is what might have been anticipated from the fact of their exposure to great alternations of temperature under the condition of physical exhaustion. Mr Field, optician, Birmingham, informs me that puddlers are moreover liable to cataract, induced by the intensely bright light of the furnace.

Workforce structure

Workforce structure in the eighteenth and nineteenth centuries is recorded in various wages books, which provide an opportunity for the workforce to be studied over a long period of change. In 1796 Horsehay forge, Shropshire, had two finers in its stamping and potting forge, employing their own assistants. The shinglers and rollermen organised their work in a similar way. In 1832 Old Park, a puddling forge in Shropshire of moderate size by national standards, had two men working at the running-out fires and up to thirteen puddlers and four charcoal finers. Otherwise its structure was the same as at Horsehay, with the skilled men employing their own assistants, and wages subject to piece rates. Whereas Horsehay had three men rolling specific products – mainly different classes of bar and plates – Old Park had five principal rollermen, also engaged on specific products – different categories of bars, plates and wire rods. Moving forward to 1849, Stirchley Forge, owned like Old Park by the Botfield family and built in 1829, employed seventy-seven puddlers, two teams working the shingling hammers and six rollermen. Numerically the workforce had grown significantly, but its structure had not changed. Key stages in the process continued to be sub-contracted to master workmen who employed their own teams. In this way ironmasters continued to recognise the authority of a core group of men in the organisation of production.

In each of the three forges mentioned there were conspicuous family groups. Continuity from one generation to the next was aided by the employment of young and old in the works. Dominant families at Horsehay were Hazlehurst, Lambert, Norton and Tranter, whose younger members were documented as such. At Old Park the Tart, Swift and Lees families, all of whom can be traced back at least to the late eighteenth century, were among the prolific family groups.

Wages books largely conceal the presence of lesser-skilled workers, whose existence was usually only acknowledged when they were paid for sundry work at a day rate. The quantity of such workers emerged by the middle of the nineteenth century when detailed accounts of work in the industry were first written.

It should be obvious by now that ironmaking was a male-dominated industry. Women and children were employed on light, unskilled work. In 1796 women at Horsehay worked collectively in filling pots with stamped iron, and it continued to be women's work until the practice of re-heating iron in pots was discontinued there in about 1830. Thereafter women did not work in forges, although they were employed in gangs in the preparation of raw materials. At the ironstone mines women were employed to pick nodules of ironstone from the mass of clay and shale brought up to the surface. The origin of such work can be traced back to the early fifteenth century when women were employed in sifting and breaking up iron ore for the bloomery at Byrkeknott, County Durham. Working on the tips was physically demanding, to the extent that few women continued with it after their mid-twenties, but it also stimulated a spirit of camaraderie among the women and allowed a certain degree of independence. The strength to carry heavy loads was put to good use as, traditionally, pit girls from Shropshire and Staffordshire left in the early summer to work in the market gardens around London. In the 1830s they could earn up to 9*s* a day carrying strawberries to Covent Garden, considerably more than they could earn on the pit banks – in 1842 Shropshire pit girls could earn no more than 1*s* 4*d* a day. A.J. Munby, a Cambridge don, took quite a fancy to the girls he encountered at Blaenavon in 1865, employed to break the ironstone

Ironmaking

61 Tip girls on strike at Dowlais, from an engraving of the 1870s.

into small lumps: 'They use heavy hammers ... lifting the hammer over their heads and bringing it down with manly skill and force. Fine strong girls they were ... they break stones thus from 6 a.m. to 6 p.m. every day, ceasing only for breakfast and dinner.' Stained with the colour of the ironstone, one woman was 'red like a red Indian, her face, which was comely, and her limbs, all glowing with ruddy sweat'.

Boys were also employed in other work preparing the raw materials, such as helping with the coke firing. In the forges and mills boys were employed from the age of eight upwards. They were employed to catch the bars after they were passed through the mills, in straightening bars with wooden hammers, and in general fetching and carrying. The essential limiting factor in their employment was the physical strength required to perform specific tasks (*see p.51*).

The status of iron workmen

The commodity that iron workmen traded was their skill. Their property was in the broadest sense intellectual property, in which context they can be seen as part of a larger category of skilled workers in eighteenth-century Britain. These included metal workers such as cutlers, as well as textile workers, printers and artisans in other manufacturing trades. In other respects, however, iron workmen were different. The blast furnace was too large to be operated by a single person, and required men to perform different tasks working as a team. At the forge, the division into two processes of fining and hammering meant that there could not be a master craftsman responsible for the overall production. The forge was therefore always a collaborative environment. These circumstances set iron workmen apart from other metalworking trades, including those craftsmen engaged in manufacturing iron, whose investment amounted to more than just their skill. In the Midlands, makers of nails, locks, buckles, saddlers' iron and scythes were independent craftsmen leasing or renting their own workshops, plant for which was normally no more than a single hearth and a pair of bellows. Production was organised around the family unit, ensuring continuity of work from father to son, while industrial work was combined with arable farming and animal husbandry. Cutlers in South Yorkshire were similarly independent, with smithies attached to cottages, as were sickle and scythe makers in the same district, who operated small water-powered grinding wheels at their smithies. Having served an apprenticeship, a cutler could establish himself in the trade at little cost. Their rural location also allowed them to supplement their income through farming.

Acquired skills and knowledge in the iron industry were inherited by subsequent generations. Iron workmen controlled their trade insofar as they could pass on their skills to whom they chose, usually sons or other close family members. It bred a closed, male-dominated, inherently traditional and empirical culture. There were no formal apprenticeships. Instead, men were said to have been 'bred up' in the trade and treated status and skill as their birthright.

Workmen were able to trade their skills across a wide geographical area. The Lavender family, of French origin, settled in the Weald in the early sixteenth century. Subsequent generations were to be found working in the Midland and Yorkshire iron trade. James Lavender, born at Wilden in Worcestershire in the early 1760s, had a career that took him to Bradley Ironworks in Staffordshire, then to numerous small forges in Monmouthshire and Glamorgan. In 1793 Thomas Lavender was working at Upton Forge in Shropshire, and in 1807–78 Samuel, Joseph and James Lavender were all puddlers at Old Park. The closed world of the forge meant that there was little spare capacity in the pool of available workmen. Mobility of skilled men and temporary absences were serious problems faced by ironmasters, especially at small forges. Charles Lloyd, of Dolobran Forge, Powys, complained that if a workman quit or suffered illness the forge had to stop work. When the iron trade expanded rapidly at the end of the eighteenth century iron workmen found it easier to obtain positions at other local ironworks. When they needed to, ironmasters poached workmen from their rivals by promising higher wages, and by the same token sought to blacklist men who did the opposite. In 1820 G.N. Grenfell, manager at an unspecified South Wales ironworks, circulated to the region's ironmasters a list of men who had left his works over a pay dispute without giving notice. 'I beg … to request that you will not receive them into your employment as this may be an inducement for them to return to their work here. Thirteen have been convicted & sentenced, but have been rescued from gaol by their fellow workmen'.

Some ironmasters looked to new technology as an opportunity to displace traditional skill. John Bedford established an ironworks at Cefn Cribwr, Glamorgan, having previously managed a forge at Trostre, Monmouthshire. By the late 1780s he was proposing to introduce a new technique of refining iron in a reverberatory furnace. Bedford thought that he could bypass established forgemen and recruit a local unskilled workforce. The anticipation was that the use of mineral fuel would simplify the refining process, with the result that he would no longer be at the mercy of a skilled elite.

Richard Crawshay had been attracted to Cort's puddling process for a similar reason. Hope evaporated with the arrival at Cyfarthfa of Cort's workmen, themselves bred up in the traditions of the iron trade.

Puddling emerged from the existing repertoire of forgemen's techniques. In the already established ironworking regions there was a long period of transition between the introduction of puddling and the demise of older techniques. Men brought up with puddling worked alongside older men who still valued the practice of granulating partially refined iron and re-heating it in pots. A protracted period of change militated against a revolution in forgemen's culture. There is no evidence that skilled men felt threatened by new techniques. Industrial 'spies' like Gabriel Jars after 1763, or Eric Svedenstierna in 1802–03, seem to have experienced little difficulty gaining access to British ironworks, where they found men willing to talk. But the ability to describe a process in no way signalled the capacity to perform it. Forgemen could be confident that the mysteries of their craft could not be stolen by a man with a notebook.

Relations between masters and workmen continued to be regulated by well-established customs. By the mid-nineteenth century men still traded skill as their property and passed their skills to the next generation without formal apprenticeships. Skilled work was still paid for by the ton, and an allowance of beer was still given to the men to distribute among their assistants. At the Level Ironworks, Staffordshire, in the 1840s, forgemen were allowed 6 quarts of ale per week, mill men 7½ quarts. Liberal allowances were not always approved by managers and ironmasters, who were concerned to maintain workplace discipline at all times. It was one of many potential flashpoints, one of the unwritten rules governing relations between ironmasters and men that generated tension. Where workmen had considerable authority in the technology of iron production, it was ironmasters who controlled the economy of the industry. Ironmasters had an interest in lowering production costs by means of wage control and improved performance, and by the same token workmen had an interest in resisting any erosion of their status. Only with difficulty, therefore, did ironmasters succeed in breaking traditional working practices. John Gibbons, who acknowledged that management of an iron ironworks was fit only for men of strong character, conceded that coercion of iron workmen was futile and that new methods should always be introduced in a conciliatory manner.

Richard Crawshay's bullying of workmen in the development of puddling is legendary. Other less formidable men found greater difficulty in putting their foot down. One of the incentives to stimulate production was to offer a bounty of one guinea if the output of furnace and forge exceeded a certain limit. In the 1790s 50 tons was a challenging target for the weekly output of a blast furnace, when the Dowlais furnaces produced 30–35 tons a week. New and larger blast furnaces at Cyfarthfa, however, saw a rapid increase in furnace capacity, to the extent that the target was routinely exceeded, but attempts to do away with the traditional bounty were protracted and difficult. Puddlers bounty of two guineas was offered to workmen at Old Park in Shropshire in the 1830s if they could puddle more than 20 tons of best iron in a four-weekly period.

Another method, tried unsuccessfully in the 1830s, was to bind the puddlers to a minimum yield. At its root, such efforts were intended to improve time discipline in the workplace when in many other industries the pace of work was regulated by machines. There were two inherent problems with the imposition of a minimum yield. First, it subtly diminished the puddlers' sense of technical authority. Second, the line separating good from inferior quality was negotiable. Puddlers found themselves in a difficult situation. Working iron to the highest quality required time, but if they worked it too long and were unlikely to make their yield, they were tempted to work it for a shorter time and risk a loss in quality. When the new regime was introduced at Horsehay in the 1830s it proved unmanageable. Quality declined, setting off a chain of blame from iron merchant to rolling mill, mill to puddling forge, forge to blast furnace and furnace to the mines, all complaining of having to work with inferior material. The classic strategy for improving the quality of the yield was simply to cheat. If they could do it without being seen, puddlers added all manner of iron such as tools or scrap to the furnace.

Welfare, accommodation and social life of workmen

Responsibility for the welfare of iron workmen was principally fulfilled by means of a works doctor and provision of accommodation. Most ironworks retained the services of a surgeon to attend to injuries in the workplace. In South Wales the custom was to pay the surgeon 1½d in the pound (0.6 per cent) on all wages earned by the men. Accidents in the works usually resulted in either burning or injury to limbs by becoming trapped in machinery. For the workmen, the arrangement was not always satisfactory. Complaints were made in 1819 that an Old Park workman had been burned on a Saturday night but, despite pleas for him to receive urgent treatment, the surgeon did not attend him until the Monday. Another Surgeon, William Edwards of Coalbrookdale, was dismissed from Old Park in 1825 for his negligence.

One of the strangest of fatal accidents occurred at Dowlais in 1793, as reported by Robert Thompson, the works manager: 'Old Edward Maddy, his wife, and another old Man, found dead in their house under the Bridge House in the Old Furnace, Suffocated as is supposed … by the Damp coming thro' the Air Holes of the Furnace into their House'. It is a reminder that dwellings were often built within an ironworks, particularly in the eighteenth century, and not just in separate uniform rows outside of the works. There are many other known examples of this practice. At Clydach, Monmouthshire, a dwelling was built beneath the charging house of one of the blast furnaces, 'a hot, smoky, dusty and noisy hole', and 'within ten feet of the window was a large waterwheel which gave motion to a very noisy blast-making machine'. At Coalbrookdale there were dwellings close to the old furnace at the Upper Works. At the nearby Upper Forge an engine house, disused by 1787, an adjacent disused malthouse, a seventeenth-century gentry house and a coach house and laundry were all converted to tenements by the end of the eighteenth century, using the waterwheel tail race as a sewer. More people lived than worked at the Upper Forge, in what became a small domestic enclave. At Blaenavon, established in 1789, the earliest purpose-built houses, Stack Square, formed an ordered block facing the blast furnaces. On the north-west side of the works a tramroad viaduct had its arches infilled and converted to makeshift dwellings, but they were short-lived as the viaduct and its dwellings were gradually buried beneath encroaching spoil tips.

The provision of housing for skilled workmen was a long-established tradition by the eighteenth century. Inventories of seventeenth-century ironworks indicate the common practice of providing accommodation for the clerk and workmen. A survey undertaken of ironworks on Crown land in the Forest of Dean in 1635, however, reveals dwellings of various standards, some of which were built by opportunist local landowners. At Parkend forge were houses for finer, hammerman and charcoal keeper, a 'little house' adjoining

Ironmaking

62 Engine Row, Blaenavon, was built 1789–92 next to the ironworks and provided the best standard of workmen's houses at Blaenavon. At the right end of the row was the original truck shop. (Crown copyright: RCAHMW)

63 Houses built beneath the arches of a tramroad viaduct at Blaenavon, as viewed by Richard Colt-Hoare in 1801, a good example of the makeshift dwellings that were common in ironworking districts. The houses, and the viaduct itself, were eventually buried beneath encroaching spoil tips.

the forge for the forge carpenter and two cottages for the other finers. Cannop furnace had a house for the clerk, another for the founder and a small cottage built by workmen and inhabited by the filler. Lydbrook furnace also had a house for the founder and three small thatched cottages built by a neighbouring farmer.

As the iron industry expanded rapidly in the late eighteenth century, ironmasters at first continued this tradition. In the remote Glamorgan and Monmouthshire uplands they had little choice, as there was rarely an existing settlement of any consequence. Housing stock was one of the principal assets of the ironmaking concerns. At Butetown near Rhymney is part of an unfinished model village built 1802–04 and based on Lowther in Cumbria. It

64 Butetown, Rhymney, one of three rows of houses built 1802−04 for workmen at Union, later Rhymney, Ironworks. Palladian in its overall composition, the layout was influenced by James Adams' Lowther village, Cumbria. (© author)

incorporated small houses for families and barracks for single men. In 1813 Hirwaun had 116 workmen's dwellings, Clydach in Monmouthshire 147. The same pattern was evident elsewhere. In the early nineteenth century 450 workmen's houses were constructed at Wilsontown Ironworks near Edinburgh on an upland site described in 1803 as 'hilly and unfertile'. Butterley Ironworks, built by Benjamin Outram (1764–1805) from 1791, was located close to the mineral wealth of the hitherto thinly populated Erewash valley near Alfreton, Derbyshire. Company housing, by 1813 amounting to 100 workmen's cottages and six houses for agents, and by 1830 some 300 cottages, was a necessity. They included the worker villages of Codnor Park and Ironville, typically not planned settlements but piecemeal developments. Most houses had two rooms in each storey, with a garden for growing vegetables, and a back yard with coal store and privy, usually shared. The standard exceeded that of cottages for agricultural labourers. In 1797 Outram offered workmen 'house, garden and land to keep a cow on easy rents', treating the houses as a necessary inducement. Rent was only 2s a week. If a workmen left the company he could still keep the cottage, but at 10s a week.

When the Furness ironmasters built their ironworks in Argyll in the mid-eighteenth century they too were faced with the necessity of constructing a settlement to house the workforce in an otherwise thinly populated district. Surviving workmen's houses at Bonawe show piecemeal development of an ironworking settlement from the mid-eighteenth to the early nineteenth century. There is also evidence of run-rig, or ploughing, from the cultivation by workmen of cereals and vegetables in traditional strips. Workmen could also obtain grazing rights for cattle on company land. A subsidiary industry at Bonawe, largely domestic and female, was the spinning of wool into yarn, although the ironmaster George Knott (d.1784) complained in 1783 that the industry had acquired a dynamic of its own and was attracting employment from people who had no connection with the ironworks.

Merthyr Tydfil was described as a village of about forty households in the 1760s, and so its ironmasters had little choice but to build company houses. Housing became

Ironmaking

a serious problem when the ironworks grew to unprecedented and unimagined sizes. Around a small upland village its three main ironworks – Plymouth, Cyfarthfa and Penydarren – grew rapidly, and its corresponding settlements mushroomed into an unregulated urban sprawl, easily the largest town in Wales by the nineteenth century. The pace of development overwhelmed the earliest house-building schemes, which were soon replaced. Visiting Merthyr in 1803 Benjamin Malkin commented that, 'The first houses that were built were only very small and simple cottages for furnace-men, forge-men, miners, and such tradesmen as were necessary to construct the required buildings, with the common labourers who were employed to assist them'. He goes on to say that ,'in some of the early, and rudely-connected streets, we frequently see the small, miserable houses taken down, and larger and very seemly ones built in their stead'.

Workmen's houses were based on rural precedents. Two-storey two- and four-room houses in Merthyr Tydfil were of a good standard compared to small rural Glamorgan cottages and the single-storey cottages of west Wales, where a substantial proportion of its workforce originated. Cheap coal allowed fireplaces to be built in both lower-storey rooms, a comparative luxury. The problems of industrial housing arose where there were dense concentrations of settlements that had inadequate provision for water supply and sanitation, and where strips of garden for workmen to grow vegetables was a less pressing concern than the desire to cram as many houses as possible into a town hemmed in by steep hillsides.

By the 1840s conditions in urban Merthyr Tydfil were as bad as Liverpool and Manchester and attracted official reports and investigative journalists, who duly spoke of the town 'abounding in fermenting and putrefying substances, equally offensive to decency and injurious to public health'. The correspondent of the *Morning Chronicle*, who visited Merthyr in 1850, found that the workmen's dwellings in the town encompassed a wide range of material worlds. The best houses:

> are of two storeys, have four small sash windows (which, by the way, are never opened), two above and one on each side the door. On the ground floor there is a roomy kitchen with a stone floor; adjoining is a small room, just large enough to contain a four-post bed, a chest of drawers, a small corner cupboard, two chairs, a window table, which usually form its contents … Above stairs are two bedrooms, one large and the other small … This is all, except perhaps a narrow cupboard cut off from the lower bedroom, and dignified with the name of 'pantry'.

The layout of such houses, usually with a steep stone winding stairs beside the fireplace, was clearly based on the layout of Welsh vernacular farmhouses. By describing the contents of such houses it was obvious that the best-paid workmen had earned good money. The houses:

> are stuffed with furniture even to superfluity; a fine mahogany eight-day clock, a showy mahogany chest of drawers, a set of mahogany chairs with solid seats, a glass-fronted cupboard for the display of china, glass and silver spoons, forming indispensable requisites for the principal room. The other apartments are equally well furnished. The habits of the women with respect of their houses, are those of cleanliness, decency, and order.

The cleanliness inside the house was often remarked upon, as at the two-room house of a Penydarren collier whose wife ran a huckster's shop selling apples, gingerbread, herrings and bacon. Elsewhere in the town people were less fortunate:

In the first house we entered, inhabited by an Irishman, there sat hovering over the fire a man swollen with dropsy; the house had no other furniture than a three-legged table, a small bench, two stools and a few utensils for cooking, such as saucepans, basins, plates and crockery. Upstairs were three beds of hay, without a single article of furniture.

After a cholera outbreak in 1849, which claimed 1,400 lives in five months, Merthyr was subject to a report by the General Board of Health. It documented the poor sanitation of the town in lurid detail – of the privies being overflowed, people having to make do on the slag tips after nightfall, of drinking water that had flowed through unregulated graveyards. Such writing was designed to shock, which it did, but, although it has helped to make Merthyr Tydfil a magnet for social historians, most of the people engaged in the iron industry in the first half of the nineteenth century lived in much smaller communities. Merthyr was teeming with life; smaller settlements in isolated districts must have seemed dull by comparison (*see p.96*).

In many of these communities ironmasters were able to dominate the local economy by the use of the truck system of payment, which was common in the Midlands and in patches in South Wales. Where ironworks were set down in hitherto isolated districts a company shop was at first a necessity. The truck system was illegal after 1831 but, if they had a mind to, ironmasters could exploit loopholes in a law that required workmen to be paid in cash. Men were paid once a month. Between those times goods were purchased in company shops, or 'Tommy shops', and the amounts deducted from the men's wages, along with their rent. Goods could be bought for cash in company shops and beer houses, but in some cases tokens were given as change that could not be tendered

65 Forge Row, part of a small isolated settlement built 1804–06 for workmen at John Knight's Varteg Forge near Blaenavon. The row was restored in 1987–88. Cwmavon House to the left was a manager's house built in around 1825 after the forge was revived. (Crown copyright: RCAHMW)

outside of the company. In practice iron companies exploited the captive market by charging inflated prices. This system was criticised on numerous grounds, including the fact that 'all habits of thrift and economy – all endeavours to provide for sickness and old age – are not only discouraged, but utterly extinguished'.

The improvidence of iron workmen was often remarked upon. A correspondent for the *Morning Chronicle* noted in a report from the Black Country in 1849 that:

> there are hundreds of families who habitually pawn their Sunday clothes every Monday morning, and take them out every Saturday evening, paying, of course, a month's interest on the loan. A strike of a couple of weeks, or the suspension for that period at a forge or mill, always suffices to crowd the pawnbrokers shops in the vicinity.

In the pawnbrokers were items of considerable value. During the insurrection in Merthyr Tydfil in 1831, workmen broke into the pawnbrokers shops and reclaimed their possessions, which included pocket watches. The seemingly excessive habits of many workmen, especially true of colliers and miners, was one of the responses to the harshness of industrial life. It is a truism that people who work hard play hard. What was true in the mid-nineteenth century was also largely true in the late eighteenth century. François and Alexandre de la Rochefoucauld, two naïve French aristocrats visiting Britain in 1785, were surprised to hear of workmen at John Wilkinson's ironworks that 'they scarcely ever live to old age: they eat and drink everything they earn and are miserable all their days. We were told not one in a hundred puts anything aside against old age.'

The public house was one of the foci of life in industrial communities. In a working environment focused upon furnaces, liquid refreshment was as necessary as it was recreational. Many iron companies owned public houses, and sometimes wages were paid there, an arrangement that obliged the workmen to purchase drink while they waited. Friendly societies and benefit clubs also met there. Drunken excess was one of the extreme responses to industrial life, and so was religion. The established church was traditionally weak in newly industrialised areas, and especially in hitherto sparsely populated districts, which allowed the various non-conformist denominations to dominate. The chapel catered for the higher aspirations of the working classes, providing status in a self-contained social world, and opportunities to develop educational and musical abilities.

Education was sometimes provided by iron companies. Early examples include the Coalbrookdale Co., who established charity schools in the late eighteenth century, and Dowlais, where the children of workmen were educated in a works school from 1828, and where, in 1852, a new school was built by Sir Charles Barry (1795–1860), architect of the Houses of Parliament. Barry also designed the Guest Memorial Reading Room and Library at Dowlais, completed posthumously in 1863 and for the use of the workmen. The autodidacts who colonised such places were a familiar type. They formed clubs like the Cyfarthfa Philosophical Society and the Coalbrookdale Literary and Scientific Institution. The Cyfarthfa brass band enjoyed a reputation far beyond South Wales. Out of such clubs came the workmen's institutes common in the coalfields in the late nineteenth and early twentieth centuries, sometimes patronised by coal and ironmasters but largely funded from subscriptions. It was the spirit of self-help and independence characteristic of Britain's heavy industries until their collapse in the twentieth century.

66 Carmel Baptist Chapel, Cefn-Coed-y-Cymmer, near Merthyr Tydfil, of 1844, a rare example in the ironworking district of a chapel retaining its mid-nineteenth-century character. (© author)

67 Coalbrookdale Literary and Scientific Institution, built in 1859 and one of the earliest surviving workmen's institutes. (© author)

10

STEEL

Bessemer and Siemens

In 1856 Henry Bessemer (1813–1898) read a paper to the mechanical section of the British Association, outlining a new method of making malleable steel from pig iron, for which no additional fuel or labour was required. The principle of the process was that molten pig iron was run into a large vessel, or converter, through which a stream of cold air was passed. The action of the air caused the decarburization of the metal, which was then cast into an ingot. The process attracted immediate and widespread interest from ironmasters, many of whom obtained licences to use it. Instant success was elusive, however, largely due to the contamination of the pig iron from phosphorus, derived either from the use of phosphoric ores, or from puddling-furnace cinders that were frequently added to the charge of blast furnaces to improve their yield. Sheffield was at the forefront of these early developments. Bessemer established a steelworks there in 1858 which began production a year later, followed closely by John Brown's Atlas Works in 1860 and Cammell's Cyclops Works in 1861. Sheffield had twenty-seven Bessemer

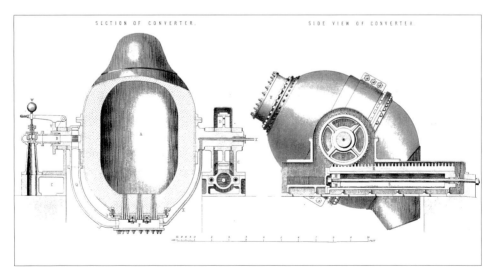

68 An early Bessemer converter. Molten iron was poured into the top and a stream of cold air was passed through air holes in the base.

converters in 1871 and thirty-six by 1878. Elsewhere, among the first works to convert successfully to steel manufacture were the Old Park Works at Wednesbury, Staffordshire, in 1864, and Dowlais in 1865.

The open-hearth furnace was patented by Frederick Siemens (1826–1904) in 1856, and was improved in a further patent of 1861 when an associated gas producer obviated the need to use solid fuel. The gas producers were entirely separate from the furnaces and yielded combustible gases from solid fuel – often of inferior grades such as slack coal, coke dust, lignite or peat – in a similar manner to a gas retort. The furnaces worked on a regenerative principle whereby two chambers, known as regenerators, were constructed of checkerwork bricks in order to build up a large surface area. Exhaust gases were passed from the furnace through one of the chambers in order to build up heat, and then the draught was reversed, so that the gas now passed through the heated chamber and entered the hearth at a high temperature where it combusted. Meanwhile the products of combustion were passed through the other 'cold' chamber. By continuously altering the direction of the draught, the furnace could be maintained at a high temperature of 1500–1800C.

Early gas-fired open-hearth furnaces were used for glass making, the melting of crucible steel, and were occasionally adapted as puddling furnaces. A further technical advance was achieved by Pierre and Emil Martin of Sireuil, France, who were able to make malleable steel by melting cast and wrought iron in a bath of molten pig iron in the open-hearth furnace. The Siemens-Martin method was a cheap way of making malleable steel using a bath of pig iron into which all manner of scrap, notably iron or steel rails, were melted. In 1866 William and Frederick Siemens established their Sample Steelworks in Birmingham in order, as its name suggests, to experiment with the new process. William Siemens then built a steelworks at Landore, near Swansea, in 1868–69, and in 1868 the LNWR Crewe Works adopted the open-hearth process for manufacturing rails.

Improvements to the puddling process

The potential of the Bessemer and Siemens innovations was not lost on ironmasters. Both allowed a significant reduction in fuel consumption and bypassed the need to have a refining stage based on manual dexterity. Both processes cast a fresh perspective on puddling as a cumbersome process, time consuming, wasteful of fuel and, as was argued, wasteful of iron as well. Whereas during the nineteenth century the capacity of individual blast furnaces steadily increased, the puddling furnace was no larger than it had been in 1800 and an increase in output could only be achieved by building more furnaces and hiring more puddlers. But there were other considerations in the appeal of Bessemer and Siemens, ostensibly intended to improve the welfare of the workmen. As has already been described, the physical strength, and the capacity to endure intense heat and the brightness of the flame, shortened the active life of the puddler. Exposure to the heat of the furnaces was exacerbated during hot summers, as in 1871, when it was reported in the *Iron Trade Circular* that some works suspended production, 'for not even the trained English puddler can stand the fearful toil'. It is debatable, however, whether 'masters and men' were both 'anxiously watching the advent of a mechanical contrivance which will supersede such exhausting work … and no object can be of greater importance to the trade than … the abolition of this well-known evil'.

Several improvements to the puddling furnace were attempted, with varying degrees of success, to compete with open-hearth and Bessemer steel. Karl Siemens (1829–1906) built gas puddling and heating furnaces at Castle Works in Shropshire in 1871 for the Birmingham screw manufacturer Nettlefold & Chamberlain. The works had twenty-seven such furnaces by 1877. Double puddling furnaces were in use by the early 1860s, whereby the furnace had a double-sized bed with opposing doors, allowing two puddlers to work it simultaneously, with a consequent fuel saving. Other attempts were made to recycle the waste heat of blast furnaces for the puddling furnaces.

The first half of the 1860s also saw serious efforts to introduce mechanical puddling. A rotating furnace was under experimentation at Dowlais in 1865, and puddling furnaces where the iron was stirred mechanically were tried in Shropshire and Staffordshire ironworks, and at Victoria Works in Derby. Stirring the iron mechanically accelerated the process, although it still relied on an experienced workman to decide when the charge was ready for the hammer. At Wombridge Forge in Shropshire Henry Bennett claimed in 1864 that mechanical puddling allowed a significant decrease in fuel costs and increase in output. An average yield per shift was 28¾cwt, compared with 22¼cwt from a conventional puddling furnace. A single furnace was said to require 28cwt of coal but a double furnace required only 17cwt.

According to John Percy, none of the mechanical puddling systems had achieved any commercial success by 1864, despite initial optimism and the fact that 'our ironmasters would be only too glad to dispense with the manipulation of puddlers, with whom so many difficulties have from time to time arisen'. The most significant advance in puddling in the 1860s was made in the Sheffield steel industry. John Brown's Atlas Forge, Cammell & Co.'s Cyclops Forge and Thomas Firth's Whittington Forge in Derbyshire all successfully employed puddling furnaces for steel making, increasing their output and expanding their market into the engineering trades. To manufacture puddled steel, iron was removed from the puddling furnace before complete decarburization had taken place, and was then melted in a crucible, but the judgement needed was a fine one, and the difficult process was never widely adopted.

A little more success came in the 1870s – between 1865 and 1875, 389 patents were awarded for improved puddling furnaces. The Casson-Dormy mechanical puddling furnace was pioneered at some of the larger Staffordshire works such as Round Oak, Chillington, and the Darlaston Iron & Steel Co.'s works at Darlaston and King's Hill. One of its inventors, Edward Smith-Casson, was manager of Round Oak. The Howatson patent puddling furnace used waste heat to supply hot air to the furnace, thereby accelerating the process and decreasing fuel costs, and also underwent trials at Round Oak in 1871. The Iron and Steel Institute sent a commission to the USA in 1871, which returned with a favourable report on the puddling furnaces used by Samuel Danks of the Cincinnati Railway Ironworks, Ohio, where iron was stirred mechanically. Subsequently some Danks puddling furnaces were erected in Britain, mainly north-east England, but totalling only seventy-four by 1873.

Mechanical contrivances never succeeded in replacing the puddler. They could work only to a predetermined cycle and, as there was no scientific control over the materials charged into the furnace, it was not possible to construct a machine capable of making the crucial judgements that would supersede the experienced eye. Interest in the mechanisation of puddling dwindled in the later 1870s when it became clear that embracing new technology would be more profitable than improving old technology. But there was not necessarily a consensus within the trade as to the desirability of continuing with puddling. In part this attitude depended upon the degree to which puddling was embedded within the local ironworking culture. In 1881 J.S. Jeans, secretary of the Iron and Steel Institute and a Cleveland engineer, condemned the traditional puddling furnace as 'crude, barbarous and wasteful'. Seven years later, at a meeting of the South Staffordshire Institute of Iron & Steelworks Managers, 'a round of applause' followed the suggestion that the puddling furnace still had a future before it.

Decline of the wrought-iron industry

The prodigious growth of the steel industry, at the expense of wrought iron, is well known. But as the iron industry was not a homogenous national industry, so the decline of the wrought-iron industry was characterised by its regional differences. To cite steel as the reason for the collapse of the iron industry is misleading. Steel was found to be superior to wrought iron for most applications, but the rate at which its superiority was established varied. Most studies of the growth of the steel industry have focused upon high-output products at large-scale works. These have shown that steel was produced commercially from the 1860s, and that steel rails were found to be more durable than wrought-iron rails and began superseding them in the 1870s. Sheffield steelmakers were quick to adopt steel for rails, having long appreciated that steel would be more durable than wrought iron if it could be produced on a large enough scale. In 1873 250,000 tons of steel rails were rolled in the city, of which the largest producer, Cyclops Works, supplied railway companies in Britain, the USA, Canada and South America. Sheffield steelmakers also rolled armour-plated steel, having already established a lead in that field in the mid-nineteenth century.

For marine engineering, steel ship plates triumphed later. Open-hearth steel for ship plates was first purchased by the Admiralty from Siemens' Landore Works in 1875. In 1880 15 per cent net tons of ships were built with steel plate, but by 1888 the figure had risen to 91 per cent. Consett Ironworks could produce 1,700 tons of finished iron per week, mainly ship plates, by the mid-1870s, but by 1886 over half of Consett's plate production was steel. It was also during the 1880s that steel replaced iron for tinplate and boiler-plate manufacture.

In the older markets for wrought iron – merchant bar for manufacturing tools, locks and other hardware products, and wire for manufacturing screws and nails – steel did not triumph until after 1900. For certain uses where resistance to corrosion was important, like chains, cable, and ships' nuts and bolts, wrought iron was still preferred up to 1914. Nevertheless, the Black Country wrought-iron trade declined significantly in the last quarter of the nineteenth century. One reason was conversion to steel manufacture – by 1883 one fifth of Black Country iron forges had diversified to roll steel billets, serving a different market to wrought iron. The most important reason was competition from Europe. From the 1870s Belgian bar, sheet and wire was sold into the Midlands below the price offered by Midland works, followed by the rise of cheap German nails and wire in the 1880s.

Other factors affected the performance of the iron industry, although whether they were instrumental in determining its long-term decline is debatable. The issue most often on the lips of ironmasters was excessive wage demands and the growing influence of trades unions. Many ironmasters and managers had an instinctive aversion to unions because they threatened their paternalistic self-image. They therefore resented the creation of institutions set up to oppose them.

Strategies for dealing with unions had often been heavy-handed. In South Wales, iron and coal masters had been alarmed at the growth of unions as early as 1831. 'The Miners Union has a most mischievous tendency', according to George Kenrick Smith, of the Varteg Ironworks near Pontypool. 'If the men are suffered to establish their union upon a regular system nothing but confusion & violence can be expected, and it is better that the evil should be crushed in the bud.' The establishment of a union among puddlers at Cyfarthfa was one of the factors in the insurrection in Merthyr Tydfil in 1831, sparked by the sacking of eighty-four puddlers, when it was said that William Thomas Williams, a thirty-two-year-old puddler, first carried the red flag as a socialist banner. The 1870s iron industry also suffered for its poor industrial relations. Robert Thompson Crawshay of Cyfarthfa locked the workmen out during a strike in 1875, and the works remained closed until his death in 1879, leaving his sons to reconstruct the works and repair the long-term damage to the family business.

Staffordshire, by contrast, benefited from the stability of long-term institutions for settling wage disputes. The 'Thorneycroft' scale, instituted in 1848, regulated wages at works producing high-quality Staffordshire marked bars. It emerged from the long tradition of equalisation of pay rates that had begun at the quarterly meetings. Other collective bargaining strategies were used in Staffordshire in the 1870s and 1880s. Wage rates were negotiated through conciliation boards made up of an equal number of employers' and trades union representatives, plus an independent arbitrator. A sliding scale tied workmen's wages to the market price for iron, and worked once the levels had been agreed. In 1874 the 'Derby scale' was introduced, whereby wages could vary between a maximum of 12s 6d per ton and a minimum of 9s 6d. Wage bargaining worked in the employers' favour by giving them flexibility over pay rates. By offering a compromise it imposed responsibilities on the trades unions to maintain production at all times and thereby reduce any losses incurred by militant action. It brought stability to the region's industry but could not stop its decline in the face of international competition.

Outsiders also singled out the conservative disposition of much of the iron trade, particularly in the long-established districts. In 1864 John Percy conceded that there were many enlightened and well-educated practical men in the British iron industry 'but many of the old stamp are still to be met with, particularly in the Midland districts'. It was said that past success had bred complacency, 'produced indifference to progress, and engendered an obstinate spirit of conservatism' that was untenable in an industry facing ever more fierce competition.

The reality was more complicated. In the Midlands and South Wales, conditions that had seen ironworks established in the late eighteenth century, based upon exploitation of local iron ore and coal reserves, had changed. Iron companies had from the mid-nineteenth century diversified into the sale of coal and the manufacture of bricks and tiles. Clay could be won easily from local coal measures. Bricks and tiles had a wide market after 1850, especially in places like the Midlands where the Victorian building boom could not draw upon cheap local building stone. Brick makers were paid at a lower rate than iron workmen, and brick-making processes and machinery were less complex and tied up less capital than ironworking. This suggests that the manufacture of bricks and tiles could turn a larger profit with less investment than working iron. Ketley and

Madeley Wood, both ironmaking concerns established in the East Shropshire coalfield in 1756, found that one hundred years later bricks and coal were more profitable than iron. The market for coal could be met with comparatively little additional investment. It is reasonable to question, therefore, whether the easier profits to be made from these industries discouraged long-term thinking in the iron trade.

A similar picture emerges from South Wales. Here, however, there was less emphasis on the clay industries because there was a good local supply of building stone. In Merthyr Tydfil the last of the major ironworks to be established, at Penydarren, was the least favourably positioned in terms of its mineral reserves. It closed in 1859, when it was purchased by the Dowlais Iron Co. for £60,000, not for the prospect of increasing its iron output, but to acquire the Penydarren collieries for the sale coal market. Cyfarthfa, with its vast mineral reserves obtained when the renewal of its lease became active in 1864, opened Castle Pit in 1865 specifically for the sale coal market. The other major ironworks, Plymouth, closed in 1880, but its collieries were kept going, the company simply switching to coal instead of iron.

By the mid-nineteenth century, when minerals leases were expiring and renewal needed to be negotiated, ironmasters found themselves in a quandary. Only with reluctance did William Crawshay II renew the lease of the Cyfarthfa Ironworks. Like many other ironmasters he had to weigh the potential profits from a capital-intensive industry against the prospect that, having made a substantial profit already from the iron industry he should do what his grandfather did, that is to invest in the industries of the future. In addition the descendants of the early nineteenth-century ironmasters had risen from trade to gentry status, a trend already described. When Thomas Botfield leased an estate at Old Park in Shropshire in 1790 it was the culmination of a long career in the iron and coal industries. The family business was carried on by his three sons, of whom William was the last to die, in 1850, having managed the Old Park side of the family interests for nearly half a century. His nephew, Beriah Botfield III (1807–1863), was the sole inheritor of the business but had lived his life away from trade. Resident at Norton Hall, Northamptonshire, he was a land owner, scholar and MP, with a different set of expectations to his grandfather. Although he was widely perceived to have 'failed' to renegotiate the lease of the Old Park Estate in 1856, his refusal to meet the terms of the landlord, Robert Cheney, was probably correct. The landlord became one of the major shareholders in a new Old Park Co. but his assessment of the potential for future profit was unrealistic. With depleted reserves of ironstone and coal, the best years had passed and the company failed in 1872, only to be sold to new owners who concentrated their business on the coal trade.

Many of the older ironworks did convert to steel manufacture. Dowlais is the outstanding example. After the death of Josiah John Guest in 1852, management of the works was left in the hands of trustees, who appointed the forward-thinking William Menelaus as their general manager. Although it embraced new technology there were looming problems for the site at the heads of the valleys. Local ironstone reserves had diminished, and the local ironstone, as in most of South Wales, contained phosphorus which rendered Bessemer steel brittle. Dowlais, in partnership with the Consett Iron Co. and Alfred Krupp of Germany, purchased hematite mines in Spain in 1873, a factor that eventually drew the South Wales steel industry to the coast. Other companies based near the coast, like Bolckow, Vaughan & Co., of Middlesborough, flourished at the expense of their less favourably situated competitors. The Dowlais Iron Co. built its East Moors steelworks at Cardiff in 1891, and thereafter it became the company's principal works.

Ironmaking

In Cumbria, the discovery in 1850 of a rich source of hematite was behind the success of the Barrow Hematite Steel Co. but, like other Cumbrian ironworks, it was originally established to smelt pig iron, with coal imported from County Durham, and exploiting the favourable coastal location. The first furnaces at Barrow began production in 1857. After the owners amalgamated with the Furness Railway Co. to form the Barrow Hematite Steel Co. in 1866 there was rapid expansion in steel making, and it had fourteen blast furnaces and eighteen Bessemer converters in 1872. Other developments in the region had begun in the 1850s. The Workington Hematite Iron Co. began production in 1858. The West Cumberland Haematite Co., which began production in 1862, had thirty-two puddling furnaces by 1865 and forty by 1872, when it converted to steel production, raising £600,000 to build a new steel plant with four Bessemer converters. The Moss Bay and Derwent Ironworks were founded in 1872 and 1875 respectively. Moss Bay originally had a puddling forge and mill but both works converted to steel manufacture less than a decade after they were founded.

Ironically the problem of phosphoric ores was solved in South Wales. Sidney Gilchrist Thomas (1850–1885) undertook experiments with his cousin at Blaenavon, where he eliminated the phosphorus by lining the converter with bricks made of dolomitic limestone and a fireproof tar mixture. This became known as the 'basic' Bessemer process, because a chemically basic lining was used, distinct from the original 'acid' Bessemer process. Blaenavon built two Bessemer converters in 1881, and enjoyed the privilege of using the patent process without paying a royalty, but it was short-lived as the steel-making plant was shut down in about 1900. The process benefited the British steel industry less than it did the German and American industry. It has been remarked that the real beneficiaries of the 'basic' process were Andrew Carnegie in Pittsburgh and Alfred Krupp in Essen.

Steel was only one factor in the demise of the iron industry, albeit an important one. There is no national stereotype into which the decline of the iron industry can

69 Barrow Hematite Steelworks in the late 1860s, by then one of the most modern plants in Britain. The blast furnaces have a substructure of cast iron columns, rather than the earlier furnaces with masonry bases. Raw materials were unloaded from railway wagons and then hauled up an inclined ramp to the charging level.

Steel

70 Barrow Hematite Steelworks in 1872. On the left is a row of twelve blast furnaces, with steep ramps for conveying raw materials, and engine houses on the left; on the right are the rolling mills.

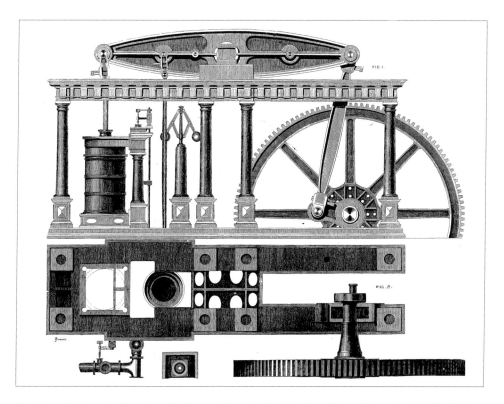

71 Ground plan and elevation of a Barrow Hematite Steelworks rolling mill engine, built by Hick, Hargreaves & Co. of Bolton.

Ironmaking

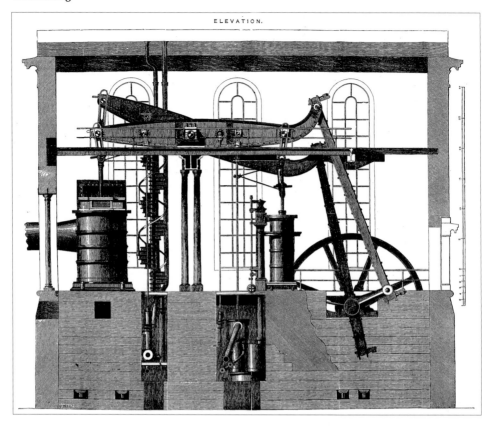

72 Double beam blowing engine at Carnforth Hematite Ironworks, by Sothwell & Co. of Bolton.

be fitted. Older ironworks, sited in the coalfields to exploit local mineral reserves, were reaching the end of their natural lives and were finished off by corporate euthanasia. Investment in steelmaking was attractive to some firms but for many, especially those for whom a conversion to steel would have necessitated penetrating new markets, it was not a viable option. The growth of the iron and steel industries of Europe and the USA threatened Britain's previous technological supremacy, and offered still more commercial competition.

The steel industry fundamentally altered the technological and entrepreneurial culture of the ferrous industries. Bessemer and Siemens were mechanical processes that no longer relied upon the experienced judgement of a workman. The technology was therefore quick and easy to disseminate. Technical supremacy could only be short-lived in an environment where rival firms could replicate any mechanical process with sufficient investment. This accounts for the rapid advance of the USA and German steel industries from the late nineteenth century. It follows that steel became a capital-intensive, and therefore high-risk, industry, and a far cry from the ironmasters at their quarterly meetings looking for their next quarterly make.

11

ARCHAEOLOGY AND CONSERVATION

For an industry that made an enormous environmental and social impact from little more than two centuries ago, comparatively little evidence has survived of the iron industry. Its sheer scale has been one of the principal factors in the transformation of its industrial landscapes, for redevelopment or re-landscaping what were once characterised as blighted or scarred places. In most cases decline of heavy industry has been a painful experience, a recipe for unemployment, general economic depression, the loss of local confidence, alleviated only by the smaller-scale light industries built, usually, on reclaimed land. Large-scale industrial landscapes of the eighteenth and early nineteenth century came with sizeable settlements which, in the post-war period, were in large part deemed unfit for habitation, sometimes against the wishes of residents, and were cleared for modern social housing. An iron industry landscape created in the nineteenth century therefore is now usually a curious combination of the old and the new. The extreme example of this is perhaps the East Shropshire coalfield, mostly now Telford, a new town on an old landscape. The Merry Hill shopping centre near Dudley stands on the site of Earl Dudley's famous Round Oak Ironworks: the archaeology of production has been superseded by the archaeology of consumerism, a fitting commentary for our times.

There was a period of forgetting before the remains of the iron industry became archaeology. Archaeologists have rarely been in the fortunate position of historians like Keith Gale who could talk to people and watch many of the processes that he recreated in his books. Derelict ironworking landscapes were once regarded as eyesores. That they can now be regarded as places of character and beauty is made possible by our psychological distance. Compare this description of Dowlais written in the 1860s, less than two decades later than Charles Cliffe's account of the awe-inspiring pyrotechnical display of the same landscape, as quoted at the beginning of this book:

> A brief sketch of a mining district will serve to show how the surface is cut up and no small portion of the land rendered waste. There are pits, conspicuous by their tall chimneys, engine houses, and huge boilers; the large unsightly pit heaps; the working and worked out brickyards, with blast furnaces, coke ovens, and heaps of slag and other debris lying here and there upon the surface. There is a network of railways of waggonways and canals for carriage of minerals, some of them in use and others abandoned, crossing and re-crossing each other, cutting up the fields in all forms, and frequently into small irregular patches. There are also

Ironmaking

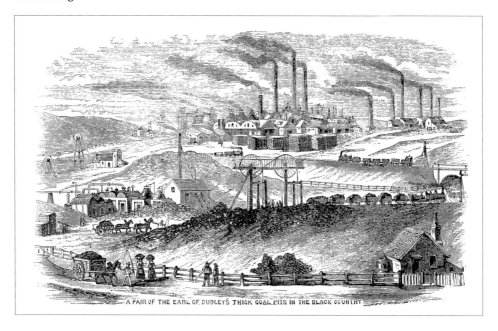

73 A Black Country industrial landscape of the 1870s, with the Round Oak Ironworks in the distance.

> the straggling rows of almost innumerable cottages for the miners, and clusters of villages, often apparently built without any regard to comfort, order, or beauty.

Few people took much notice of these places until the 1950s, and among the first people to do so were creative artists. Sat in a public house on the slopes of the Blorenge, close to the ruins of Garnddyrys forge that had closed nearly a century before, Alexander Cordell was fired up by stories of the ironworks. He used the local landscape as the setting for *Rape of the Fair Country* (1959), the first of a series of novels romanticising working-class struggle in the iron districts. No matter how devastated the landscape is, our urge to identify and evaluate the past remains undiminished, given a sense of greater urgency by the pace of change in the built environment. Having rediscovered the remains of the recent past, this chapter is concerned with how we have made sense of the archaeology of the iron industry.

If any theme has dominated the archaeological study of the iron industry it is the development of technology. There are good reasons for this – the archaeology usually speaks directly of technology and the intellectual development of the subject has been led by investigation of sites of the early modern period, for which period technology would otherwise be uncertain. Nevertheless, it is as well to remember that what, to archaeologists, is essentially obsolete technology, was perceived very differently in contemporary minds. These archaeological sites were places of work when work was the most influential component of self-identity, and when the workplace represented a hierarchical structure within which people understood their place in the immediate world. Ironworks embodied a self-contained culture. The sight of a working blast furnace might have stimulated a variety of responses – as a place of opportunity or a place of memories, liberation from country life, or a place of personal tragedy – just as an abandoned ironworks could evoke bitter memories for those who knew what it used to be. If ironworks and related sites can be understood at this personal level, it reduces the unfortunate tendency in industrial archaeology for over-emphasis on the ranking of sites according to technological 'firsts'. The root of this thinking is of course the glorification of the industrial past that has already been mentioned. Any ironworks, large or small, embodies a myriad of personal experiences, and of skills that have been entirely lost to the world. The archaeology of the iron industry also has a much broader relevance. Glorification of the industry's achievements might be one way of understanding it, but the archaeology can be read in terms of both triumph and environmental degradation. Iron offers a case study of how industry uses a place up and then moves on, indifferent to the social and topographical scars it leaves behind. The archaeology of ironmaking is a commentary upon the nature of industrialised society.

The role of the archaeologist is to explain the remains of the iron industry, to understand the reasons for survival and loss, and to justify its conservation. In conservation, practical constraints have usually intervened. Remains of the iron industry are not always easy to adapt to other uses. Small sites like the charcoal blast furnaces described below have been conserved for display as ancient monuments. Larger ironworks self-evidently require larger resources and the practicalities of conserving such sites are made more difficult. Blists Hill in Shropshire and Blaenavon are the only conserved ironworks that could reasonably be described as large sites. Where they have survived, engine houses and other ancillary buildings have proved suitable for conversion to houses, offices or other uses.

When it comes to the wider built environment, conservation has been more controversial. Understanding the significance of working-class housing has been used as an argument for its preservation, but it has also been used as an argument for its demolition. The collective memory that heritage tries to tap is not always benign. The problem has been that post-war clearance of the worst class of dwellings was accompanied by demolition of much of the better quality of workmen's houses that would be adaptable to modern living, and contribute much needed local character and identity. Clearance continued as late as the 1980s. Plymouth Ironworks near Merthyr Tydfil had some of the most innovative and well-preserved iron-industry housing, like the Triangle demolished in 1979 and the Squares in Abercanaid, needlessly demolished in 1987. In the process, the locality lost part of its unique culture.

Survival of sites of both the charcoal and coke periods of the iron industry share a number of common factors. Isolation accounts for the best-preserved ironworking sites, because there has been less pressure for redevelopment. Longevity has also been conditioned by the solidity of the original structures. This is why the masonry encased blast furnaces common until the mid-nineteenth century have survived but those of the later period have not. Calcining kilns constructed in a similar way have also survived for the same reasons. Furnaces that were constructed of firebrick towers clad in iron sheets were not intended to survive and their structural ironwork was invariably removed for scrap when a works closed. Later ironworks were also built on a larger scale, leading to more pressure for redevelopment of redundant sites. The same general conditions apply to the ironworking landscape. Those situated in outlying upland districts have survived much better than in the well-populated Midlands. Nevertheless, early mining remains in the coalfields often retain substantial coal reserves that are easy and cheap to exploit by open-cast techniques, a practice that completely obliterates evidence of earlier activity.

Ironmaking

74 The Triangle at Pentrebach, Merthyr Tydfil, built for workmen at the Plymouth Ironworks, c.1840. The houses were demolished in 1979. (© David Anderson)

The charcoal iron industry

Among the earliest upstanding blast furnace remains, of the early seventeenth century, is at Trellech in Monmouthshire. This and the slightly later Abbey Tintern Furnace, which was excavated and conserved for display in 1979–80, reveal a characteristic layout. In each case the furnace is only preserved in ruinous condition, but the outlines can be seen of the blowing and casting houses, and the wheelpit and tail race. Further uphill each has remains of a charcoal store. At many sites only the blast furnace has survived, of which Rockley in South Yorkshire (later converted to coke smelting) and Charlcotte in Shropshire are among the most complete. In Wales only one charcoal blast furnace survives to its full height. Most are partial survivals, like Caerphilly, Carmarthen and Melincourt in South Wales and Dol-gun in North Wales. The majority remain under-investigated – Abercarn blast furnace in the Monmouthshire valleys was only discovered in 1993.

Furnace complex, storage buildings and housing are the principal elements found in ironworking sites of the period, and always constructed close to the rivers and streams that powered the waterwheels. This type of layout remained standard in the eighteenth century, to which period belong the three best-preserved charcoal blast furnace sites in Britain – at Duddon in Cumbria, Dyfi in Ceredigion and Bonawe in Argyll. All are now conserved ancient monuments, and all were associated with Furness ironmasters. Dyfi was built in around 1755 by Vernon, Kendall & Co in a region with a good water and charcoal supply and good communications by sea. Iron ore was imported from Cumbria. The furnace had a short life, having been blown out by 1810, and the site therefore underwent little alteration from its original layout. The blast furnace has a charging house behind it which is built into a natural bank, in the lower storey of which was the

Archaeology and conservation

75 Rockley blast furnace, South Yorkshire, with the casting arch to the front and blowing arch on the left. (© author)

bellows house. The water required only a short leat and short tailrace back into the Cwm Einion brook, but the surviving waterwheel is a replacement. Dyfi also retains a charcoal store on the bank above the furnace.

Bonawe is similar but it survives on a larger scale. It enjoyed a comparatively long period of use, from 1753 until 1876, although smelting was intermittent after 1850. Like Dyfi, iron ore was imported to Bonawe and so the immediate hinterland has no evidence of mineral extraction, only of charcoal burning. Although its charging house and furnace survive to full height, the casting and blowing houses survive only at foundation level. Excavation has revealed that the original bellows were replaced in the nineteenth century by blowing cylinders, but were still powered by a waterwheel (which was removed only in 1941). Above the furnace are iron ore and charcoal sheds with their original roofs. Much of the associated iron industry settlement at Bonawe can still be traced. Around the furnace site are remains of workmen's houses, and a manager's house built in 1772 but extended in the nineteenth century. Workmen's housing includes a block of former tenements begun in the eighteenth century, and a former truck shop.

Duddon was built in 1737-38 and was finally blown out in 1867. It is similar to Bonawe in its plan, with blowing house and casting house surviving only at foundation level. Excavation has revealed that the bellows were replaced in the nineteenth century by blowing cylinders, concurrent with the improvement of the water supply. Fragments of a mid-nineteenth-century waterwheel have been recovered, calculated to have been 8m in diameter, and of breast-shot type. The furnace survives to full height, although it has lost the former pyramidal roof at the charging level, which was built around the central narrower stack, which has survived. The charging house, now roofless, has a bothy for workmen in its lower storey, the kind of rudimentary dwelling already referred to at Dowlais and Clydach ironworks. At Duddon the charcoal and ore sheds, also

Ironmaking

76 Dyfi furnace, Ceredigion, with its restored waterwheel, and the bridge house, beneath which were housed the bellows. (© author)

77 A cast-iron lintel across the casting arch at Bonawe, Argyll. (© author)

78 The upper level of Bonawe furnace, showing the iron-ore shed on the left and one of the charcoal sheds further behind on the right. (© author)

now roofless, form a compact block above the furnace, although the present L-shaped charcoal store represents several phases of enlargement.

Duddon and other surviving Furness blast furnaces – Backarrow, Nibthwaite and Newland – as well as former smelting sites where storage buildings still stand – Cunsey, Leighton, Lowwood and Penny Bridge – can also be seen in a broader context. Recent surveys have revealed much of the area's ironstone mines, the woods in which charcoal was burnt, and the transport infrastructure that served them, which provides a good contrast with the case study of Blaenavon described later in this chapter. At Furness iron ore and charcoal were exploited in different areas. Most of the iron ore was extracted from Low Furness between Ulverston and Barrow-in-Furness, or on the west side of the Lake District in Eskdale. Woodland was exploited in High Furness, between Lakes Windermere and Coniston, where the majority of the furnaces and forges were built.

The earliest evidence is of surface workings which, by their very nature, are difficult to date. They form linear depressions as they follow a vein of ore, although it is also difficult to distinguish between small-scale workings and trenches dug to ascertain the extent and quality of the ore before large-scale working began. In the eighteenth century the ore was extracted in large workings known as 'sops', or was won from underground drifts or adits. These would be dug into the vein, which was exploited by a series of steps, or 'stopes', creating voids that were backfilled with waste. The ore was raised to the ground from drifts and shafts by means of horse-powered gins. However, the character of the Furness ironstone mines has been dominated by the larger nineteenth-century workings. Archaeological evidence of calcining in Furness, as elsewhere, is minimal. The outstanding but untypical example is the excavated calcining kiln at Allensford, Northumberland, built near a blast furnace in operation from the 1670s. It represents a break with the tradition of roasting the iron ore in open heaps that would continue long into the coke era.

The woodland industries have left more evidence although, again, the date of many of the features is difficult to ascertain. Exploitation of timber resources for charcoal brought on subsidiary industries that have left some archaeological traces, mainly bark for the tanning industry, potash for the woollen cloth industry, wood for bobbin making and charcoal for gunpowder manufacture. Woodlands were divided into small parcels of ground, known as 'coupes' where the wood was cut in rotation. Charcoal burning was a seasonal activity, usually undertaken between August and November. First, the ground had to be levelled, known as a charcoal platform or a pitstead, to about 7m in diameter. Wood was burned in a circular clamp in which wood was stacked to form a flat dome and sealed with turf and bracken to prevent oxygen seeping in and thus allowing it to burn too quickly. The burn lasted for up to two days, and the charcoal was ready when the smoke stopped emitting from a narrow central flue.

Archaeological evidence of charcoal burning is slight, usually being confined to the pitstead, many of which have been identified in High Furness woodlands, often by the dark nature of the soil. Little evidence has been found of charcoal burners' huts, being temporary in nature. Bark peelers also lived in huts which were more substantial if still rudimentary, having stone chimneys and turf or stone walls. In Roudsea Wood near Haverthwaite an unusually large number of huts have been identified, as well as a bark barn, situated close to a bloomery.

Forges of the charcoal period are rare survivals, although the site of large numbers of them is known in a context that suggests the great potential for archaeological fieldwork,

Ironmaking

79 A reconstructed charcoal clamp in the Caledonian Forest. (© author)

including excavation. The short-lived Stony Hazel forge, built in Furness in 1718–19, unused by 1725 and abandoned by 1743, has been the subject of three excavations, the most recent in 1985. The layout of the site can be discerned from the upstanding but ruined remains. These include the low stone walls of the forge building, inside which is the base of a finery hearth, a now roofless charcoal store, and a leat, header pond and tail race. Pitsteads in surrounding woods are evidence of charcoal burning, although the woods were exploited long after the forge ceased work.

The archaeological evidence for Stony Hazel is ambiguous. Documentary evidence shows that it was built as a bloomery. As it is the last known bloomery to have been built in Britain, its form is not necessarily representative of earlier bloomeries – it has been argued that the bloomery in seventeenth-century Cumbria derived much of its technology and equipment from finery forges in the Weald and Forest of Dean. Hematite was found in an ore bin, and even in the hearth itself, which appears to confirm the historical evidence. But at the back of the hearth is an aperture plausibly interpreted as a pig-hole. The aperture was used in a finery hearth for slowly feeding pig iron into the furnace. It remains an open question whether the forge was intended to be converted to a finery, whether it was intended to use both bloomery and finery technology, or whether the refining of pig iron at the site was only experimental. Stony Hazel is an important site, not least because its explanation is not as straightforward as historical sources imply. It demonstrates how close analysis of an ironworking site can challenge pre-conceived categories in a diverse and complex industry.

The best surviving forge of the charcoal period is probably Top Forge at Wortley, South Yorkshire. The main forge building is dated 1713, although from around 1840 it was

Archaeology and conservation

80 Top Forge, Wortley, South Yorkshire, showing a forge hammer of eighteenth-century origin, and a restored waterwheel. The hammer is of the belly helve type, lifted by a cam on the underside of the beam. (© author)

altered significantly for the production of railway axles. The forge itself is a substantial stone building, and has waterwheels against both of its long walls, in which sense it gives the general impression of how the original finery was laid out. The details are all later. It retains three waterwheels, all nineteenth-century cast-iron replacements that powered two hammers and a blowing cylinder. The two hammers are of the belly helve type, i.e. the hammers were lifted by a cam operating on the underside of the hammer beam or helve. Both were probably used for shingling and drawing the iron into bars, before alteration to hammer the axles. Four cranes of timber construction were used for moving the axles. As its use changed in the nineteenth century, so additions were made to the original building, including two cottages and a smithy and foundry. Archaeology of the nineteenth-century working at Top Forge is also of national significance, even though it was a comparatively small example of a branch of the industry that has left all too little traces.

The coke iron industry

Just as the coke iron industry began in Coalbrookdale, so did the heritage of the iron industry. Coalbrookdale and its vicinity retain the most impressive remains of the eighteenth-century iron industry, with significant remains of the nineteenth century too. Nikolaus Pevsner described the blast furnace at Coalbrookdale in 1958 as 'shockingly sordid', needing only a little money to create a worthy monument to early English industry. Efforts to preserve Abraham Darby's blast furnace at Coalbrookdale gathered pace in the 1950s at what was still, technically at least, part of the Coalbrookdale Co.'s Upper Works. In 1959 a small museum was opened by the company, commemorating the 250th anniversary of when Darby began smelting with coke. Gradually, as the company rationalised its operations, the old upper works was given over to museum and other heritage uses, forming the core of the Ironbridge Gorge Museum's Coalbrookdale site. Evidence of the iron industry at Coalbrookdale goes far wider than the blast furnace itself, and includes structures from the seventeenth to the twentieth century.

The first blast furnace at the Upper Works was built in 1658 but the surviving furnace is a complete rebuild of 1777, re-using dated cast-iron lintels from the earlier furnace (*see p.34*). Smelting continued there until around 1816. As a blast furnace site, the layout of the Upper Works was derived from its seventeenth-century origin and did not differ significantly from other charcoal furnaces of the period. Behind the blast furnace is a pond and a masonry dam from where there was access to the charging house. On the side of the furnace was a waterwheel for powering the bellows (replaced by blowing cylinders in the late eighteenth century), and a separate casting house. Neither the bellows house nor casting house have survived but, unlike the upstanding remains which have received detailed investigation, in both cases there is a high potential for future excavation, particularly given the early use of air furnaces at the site. The cast-iron lintels were dated 1658 but of that furnace, only the lintels themselves, the rubble-stone footings, and part of the masonry walls and dam behind it, survive. The rebuilding of the furnace by Abraham Darby III was not, as has sometimes been believed, directly connected with the supply of cast iron for the Iron Bridge. It was part of the general expansion of the various Coalbrookdale ironworks in the 1770s and 1780s. Another furnace, used as a store in the latter part of the nineteenth century, was a 'snapper furnace' built against the dam wall. Snapper furnaces were small blast furnaces with a yield of about 10–15 tons of pig iron per week when Shropshire blast furnaces could produce up to 30 tons. Five are known to have been built in Shropshire at the end of the eighteenth century and were intended to be used as supplementary furnaces when there was a high demand for pig iron. The Coalbrookdale snapper furnace, the only known survivor of its type, was built in the last decade of the eighteenth century, although there is no evidence that it was ever finished and blown in.

Coalbrookdale was perhaps the first ironworks to have expanded from blast furnace to a foundry and engineering works. Most of the early engineering workshops have long since been demolished but there are substantial remains of a turning mill, built in about 1780 on the site of a boring mill built in around 1722 for the boring of engine cylinders. It remained water-powered throughout its active life. After the cessation of smelting at Coalbrookdale, the works became part of a large foundry complex. The small scale of the buildings at the upper end of the works can be contrasted with the warehouses and engineering shops elsewhere on the site, which are far larger in scale and include a warehouse dated 1838, to which a cast-iron clock tower was added in 1843. The site of the second blast furnace at Coalbrookdale, erected by Abraham Darby I in 1715, is still a working foundry, and has been extended further downhill to occupy the site of the former Upper Forge pond.

The original Upper Forge at Coalbrookdale, a finery and chafery for which the earliest known date is 1668, was demolished in 1938 when the road was widened, but a later building, erected in 1776 as a slitting mill, has survived. It had a waterwheel to drive the mills but it was soon adapted to working a hammer when stamping and potting was

Archaeology and conservation

introduced at Coalbrookdale. An engine house was added in 1785, the earliest surviving engine house for a Boulton and Watt engine used in the iron industry. The Upper Forge is also the only upstanding building associated with stamping and potting. It remained in use as a forge in the early nineteenth century, probably increasingly for specialised work, but by 1838 it had been converted to a mill for animal feed. The front of the building retains the character of a mill and stables. Only at the back, inside and beneath it are there still traces of its complex earlier history. Of sites further downstream, the Middle Forge was converted to a boring mill in 1734, and the Lower Forge was a secondary forge where frying pans were manufactured from at least 1660, but neither has survived. The earliest identifiable ironworking site at Coalbrookdale is the seventeenth-century steelworks, where excavation has recently revealed evidence of a cementation furnace. Sir Basil Brooke, owner of Madeley manor, was awarded a patent for manufacturing steel in 1612. The site is between the Middle and Upper Forges, and close to a row of timber-framed cottages dated 1642. Brooke probably imported bar iron from his forges in the Forest of Dean. Shipments of steel down the River Severn are recorded in the Gloucester port books over the period 1615–80. Further downstream, on the banks of the Severn, is a castellated warehouse of 1862.

Coalbrookdale retains the character of an industrial settlement. Most of its housing is nineteenth-century, after the period of smelting, mostly of brick and including many houses with iron-framed windows, including a mill from around 1821 which has now been converted to dwellings. The blue-brick Coalbrookdale Literary and Scientific Institution was built in 1859 and was one of the earliest buildings of its type (*see p.115*).

81 The Upper Forge at Coalbrookdale. It was built in 1776 as a slitting mill and was converted for stamping-and-potting in around 1783. An engine was added in 1785, represented by the taller section on the left of the building, but ironworking ceased in the early nineteenth century. It was later converted to a mill and stables. (© author)

Ironmaking

Coalbrookdale also retains significant buildings of the eighteenth century, including two ironmasters' houses within sight of the Upper Works *(see pp.37, 100)*. Dale House was begun by Abraham Darby I but was unfinished when he died in 1717. It was enlarged in the late eighteenth century. Directly uphill is Rosehill House, built by Darby's successor, Richard Ford, in the 1720s. Both are well-preserved Georgian houses of obvious status, but remain within the industrial settlement. Most of the workmen's houses were built lower down the valley, but further up, just above Rosehill House, is the surviving Tea Kettle Row, a piecemeal development of one-and-a-half-storey houses built over the period 1735–46. This, and two other late eighteenth-century rows in Coalbrookdale – Carpenters Row and Engine Row – are remarkable survivals.

In technological terms, the next phase of development beyond the Coalbrookdale Upper Works was the introduction of steam power at the blast furnace. Bedlam is the best preserved of the new generation of coke-fired blast furnaces built in the East Shropshire coalfield in the 1750s. Built in 1757, furnaces remained in blast here until 1843 when its owners, the Madeley Wood Co., switched production entirely to a new site at Blists Hill. Bedlam has substantial remains of a Newcomen engine house, and an engine pit from which water was pumped from the adjacent River Severn for supply of the waterwheels. In its original form it had two waterwheels each turning an axletree with cams depressing the bellows, raised again by counterweights. Substantial evidence for one of these wheels and a bellows house has survived. Later, as in most furnaces of the period, the bellows were replaced by blowing cylinders. The present furnaces at Bedlam are larger nineteenth-century replacements, but it is significant that the blowing cylinders remained water-powered throughout the lifetime of the works.

82 Carpenters Row, Coalbrookdale, built in the 1780s. Originally the houses had vegetable plots in front but the sense of them was lost when the track in front of the houses was widened to a road. (© author)

Archaeology and conservation

83 Bedlam Ironworks was built in 1757. Its two blast furnaces are later, probably of the third decade of the nineteenth century. (© Ironbridge Gorge Museum Trust)

Blists Hill was built in 1832 and represents a phenomenon also seen in South Wales – of an ironworking company founded on a leasehold site who moved to a more financially advantageous freehold site (*see p.68*). The Blists Hill furnaces were blown by a steam engine at a time when availability of water power ceased to be a consideration. The earliest of the surviving engine houses is a well-preserved beam blowing engine house of 1840. The later, built in 1873, housed a vertical engine. Together, therefore, Bedlam and Blists Hill demonstrate three important phases in blast-furnace blowing technology. The three furnaces at Blists Hill retain their original furnace bases, one built in 1832 and two in 1840, but the superstructures have not survived, typical of blast furnaces of the second half of the nineteenth century which, constructed as firebrick towers clad in iron sheets, were provisional structures taken down when the furnaces were stopped. Bedlam, Blists Hill and the Coalbrookdale Upper Works were all closely associated with the foundry trade and were therefore never integrated with forges. The Madeley Wood Co. specialised in cast iron because the local Crawstone and Pennystone ores yielded pig iron that was particularly suitable for foundry use. For the same reason Blists Hill continued to smelt with cold blast until 1912.

Bersham near Wrexham is an ironworks that in many respects parallels developments in Shropshire and was also associated primarily with the foundry trade. Built in 1717, in 1721 it became one of the first blast furnaces to smelt with coke. After its acquisition by Isaac Wilkinson in 1753 the site became a foundry and engineering works at the forefront of technological development. It was inherited by the brothers John and William Wilkinson but the works did not survive the brothers' quarrels and was closed in favour of John Wilkinson's Brymbo Ironworks, built in 1792. The blast furnace at

Ironmaking

Bersham has been revealed by excavation. A surviving octagonal building from around 1775 is one of the earliest purpose-built foundries designed to draw upon multiple air furnaces, in the period before the cupola furnace was invented. Excavation has also revealed evidence of earlier air furnaces.

There are many other surviving coke blast furnaces, mostly in South Wales, and mostly of the late eighteenth or early nineteenth century. Some of the survivors – like Moira furnace in Leicestershire, built in 1803, and Whitecliff in the Forest of Dean, in blast from 1804–16 – demonstrate the seemingly arbitrary nature of archaeological survival. As archaeological monuments they are of national importance, but in the contemporary iron trade they were peripheral (notwithstanding the latter's association with David Mushet). Compare these with Dowlais Ironworks, once the largest ironworks in the world, where none of its eighteen blast furnaces have survived, and where the only surviving building directly related to iron production is a late nineteenth-century blast engine house, built long after the conversion to steel. Moira and Whitecliff were failed ventures – both lasted less than two decades – which places them in a category with other ironworks like Banwen at Dyffryn Cellwen in the Dulais valley, a site with two blast furnaces begun in 1845, but which never entered sustained production. They were unwise speculations occupying sites that were never required for redevelopment. This factor has accounted for the survival of many blast furnaces in comparatively outlying areas, and is one reason why the archaeology of the Black Country iron industry is so scarce. Another small ironworks of the coke period, Cefn Cribwr, Glamorgan, retains the blast furnace built in 1790 and a ruined cast house, with calcining kilns above the charging level.

84 The early nineteenth-century Moira Furnace, by the Ashby Canal in Leicestershire, one of the earliest surviving brick-built blast furnaces. (© author)

Archaeology and conservation

85 The only surviving building directly related to smelting at Dowlais, once the world's largest ironworks, is this late nineteenth-century engine house in uncharacteristic brick. (© author)

At many of these sites the blast furnaces and some of the ancillary buildings have survived. Clydach is one of the best preserved ironworks in South Wales, and well represents the generation of ironworks built along the heads of the valleys in the late eighteenth century. It was built in 1793 and occupies a narrow valley site well suited to water power but less well suited to the kind of large-scale expansion seen elsewhere in the nineteenth century. A second furnace was added in 1797 and a third in 1826, the latter surviving only at hearth level. Excavation revealed much of the layout of the works, including casting houses, bases of cupola furnaces and a possible air furnace. The charging house of the first furnace survives partially, and adjacent to it is the wheelpit. The constricted nature of the site explains its difficulties in the nineteenth century, when it failed to establish a specialist market but struggled to compete with the larger works. After 1813 it had a succession of owners, none of whom could make it profitable, despite adding a fourth furnace by 1839. The last three owners all went bankrupt, before final closure in 1877.

Although engine houses have survived in gutted form, a few engines escaped the scrap merchants. A beam blowing engine built in 1817 for the Netherton Ironworks stands incongruously on a roundabout at the southern end of the A38 Aston Expressway in Birmingham. At the Blists Hill Victorian Town in Ironbridge is a re-erected double beam blowing engine, nicknamed David and Samson, built for the Lilleshall Co.'s Priorslee Ironworks in 1851. A vertical blowing engine from Priorslee has been squeezed into an engine house built in 1873 at Blists Hill furnaces.

Gadlys in Aberdare (built in 1828), Ynysfach in Merthyr Tydfil (built in 1801 and enlarged in 1836) and Llynfi in Maesteg (1839) retain blast furnaces and an engine house. The engine houses are perhaps the best surviving class of building in the characteristic industrial style. Ynysfach and Llynfi engine houses are freestone buildings of pennant

Ironmaking

86 Ynysfach engine house built in 1839, and typical of the South Wales industrial style with its freestone masonry with lighter quoins and dressings. It housed a beam engine manufactured at Neath Abbey. (© author)

87 Llynfi Ironworks engine house, built in 1839 and successfully converted as part of a sports centre. (© author)

sandstone with lighter limestone dressings to round-headed openings, and have restored hipped roofs. The tall proportions were dictated by the need to accommodate a beam engine. The architectural character has strong similarities with local non-conformist chapel architecture, simply because the builder-architects who designed the engine houses were the same builders working on chapels. A parallel can be seen in Shropshire, where the 1840 engine house at Blists Hill incorporates the same motifs as the local chapels, deriving from the same builder-architect tradition, and in the Italianate engine house at Dalmellington, Ayrshire.

Similar style was incorporated into other ironworking buildings, very few of which have survived. Large-scale forges on flat sites were particularly vulnerable to clearance and have not survived. At Neath Abbey, where there are two massive furnaces of 1793, a forge and rolling mill of 1825 has survived thanks to re-use as a factory, and is in the austere manner of many industrial buildings in South Wales. It has an associated dam and watercourses that supplied the waterwheel. The engineering works at Neath Abbey survives only as a roofless shell. A related site that well represents the kind of industrial architecture favoured by the South Wales ironmasters is the Treforest tinplate works near Pontypridd. Archaeologically Treforest is an important site because it was occupied previously by a finery forge, which is listed in the national survey of the iron industry in 1715. It was converted to a tinplate works by William Crawshay II which, after a stuttering programme of building work, began production in 1835. The original building is characteristic of the Crawshay style, of stone with round-headed doorways and windows, and bullseye vents in the gables. Near to it stands a smithy, erected with the expansion of the works in around 1854, although possibly re-using the components of an earlier building. It has a cast-iron frame of uprights spanned by arched braces, which have open circles in the spandrels. This style had been used from at least 1825 when views of the interior of the Cyfarthfa forges and mills show roofs of a similar construction. The Treforest smithy is the only example of the pioneering prefabricated iron construction known to have survived in South Wales.

There is little surviving archaeology of materials preparation, partly because in many regions there was a preference for calcining and coking in open heaps. An eighteenth-century coke oven has been discovered at Bersham, and at Bedlam the site of experimental coke ovens erected in 1789 remains undisturbed. Most examples are from the second half of the nineteenth century, like the extensive remains at Tondu (rebuilt from 1854) and Cefn Cwsk (1839), near Bridgend. In England there is a group of coke ovens at Vobster Breach Colliery in north Somerset, from where coke was supplied to the Seend and Westbury ironworks in Wiltshire in the 1870s. In South Wales, Tondu, Blaenavon and Gadlys all retain banks of calcining kilns, while the group at Rosedale in north Yorkshire were constructed in the 1860s at the ironstone mines rather than the ironworks they supplied.

Some landscapes of the iron industry

Among the largest of the South Wales ironworks were Cyfarthfa and Blaenavon, both of which have a significant landscape context. At Cyfarthfa iron was smelted from 1765 to 1875 then, after a lock-out lasting four years, a steelworks was constructed with new blast furnaces that began production in 1884, but by 1902 it had been bought by GKN,

Ironmaking

88 Calcining kilns at Blaenavon Ironworks, built into a bank to facilitate charging from the top. The iron ore was drawn from the bottom, which is at the same level as the blast furnace charging platform. (Crown copyright: RCAHMW)

owner of its great rival the Dowlais ironworks. The furnaces ceased work in 1910, with a brief reprieve during the 1914–18 war. The six blast furnaces that ceased work in 1875 have survived. A sketch made in 1797 by J.M.W. Turner shows four furnaces, only one of which is still standing. Another three were built by 1824, although one of them had already been taken down before the great strike of 1875. (The Crawshays also had four blast furnaces at nearby Ynysfach Ironworks, making a total of eleven furnaces.)

The furnaces were blown with water power until the 1820s, and the rolling mills remained largely water powered throughout their working life. The works was situated just below the confluence of the Taf Fawr and Taf Fechan Rivers, and water was drawn from both sources. On the Taf Fawr a weir survives and part of the original leat can still be traced. On the Taf Fechan a leat can be followed for some distance. It crosses the river into the works by an aqueduct known as Pont-y-Cafnau, an iron tramroad bridge built in 1793 that incorporates a trough below the deck. High above this bridge another watercourse was carried on trestles, that supplied *Aeolus*, the great waterwheel for blowing the blast furnaces. The site of the Cyfarthfa forges and mills has been mostly redeveloped for light industry, but below ground a remarkable series of stone-built vaulted tailraces survives, which directed the water back into the River Taf.

This part of Merthyr Tydfil was once a dense network of small streets in enclaves known as Georgetown, Williamstown, Ynysgau and Chinatown. Clearance of the majority of its workmen's houses occurred in the post-war era to the 1970s, and even the structure of the former settlements has been lost now by housing development and a new road layout. Nevertheless, some elements of the ironworking settlement have survived, the most significant of which is Cyfarthfa Castle.

Archaeology and conservation

Cyfarthfa Castle is a remarkable building but not a very attractive one (*see fig. 59, p. 101*). A castellated mansion with rugged detail, it is nevertheless austere and unimaginative in its planning and, but for a Gothic entrance hall, has rather staid classical interiors. Its scale and its position nevertheless make it an unrivalled embodiment of capitalist power in the iron industry. According to its architect, Robert Lugar, the view from the house offered 'a truly magnificent scene, resembling the fabled Pandemonium, but upon which the eye may gaze with pleasure, and the mind derive high satisfaction, knowing that several thousand persons are there constantly employed and fed by the active spirit, powerful enterprise, and noble feeling, of the highly respected owner.' In the grounds of the castle is a large lake, which also acted as a feeder for the works, and whose construction was said to have cost almost as much as the house. Immediately below is Pandy Farm, a sort of home farm for an ironmaster dreaming of becoming a country squire.

Further downhill are remnants of former workmen's housing. Williamstown retains two rows of four-room houses built in around 1840, a good contrast to Cyfarthfa Castle. Nearby, beside the Glamorganshire Canal, are five houses at Chapel Row from around 1825, again four-room houses but among the largest of their type in Merthyr. The earliest rows are on the opposite hillside behind the ironworks but are much altered. At Gellideg is a vernacular farmhouse which was progressively extended to make a short industrial workers' terrace over the period 1765–94. Next to it stands a row of single-fronted houses, built in 1797, which had outshuts to the rear. This was the standard early Cyfarthfa house type. It comprised a single room on each floor, with an additional ground-floor sleeping room and a larder in the outshut. The best-preserved example of these houses, built in 1804 at Rhydycar for workmen at Cyfarthfa's ironstone mines, has been re-erected at the Museum of Welsh Life. The type remained common in Merthyr, and was used by the Crawshay family at Treforest Tinplate Works as late as the 1830s. Beyond Merthyr town is evidence of ironstone and coal mining, although compromised by recent road building. Transport networks have also survived in part, including the Cyfarthfa Canal, a tub-boat canal used for conveying ironstone and coal to the upper level of the works, and the tramroad that brought limestone from quarries to the north along the bank of the Taf Fechan River.

Tramroads and associated features have survived in connection with many of the South Wales works, including Hirwaun, Nantyglo, Blaenavon and Clydach, and the Penydarren tramroad that ran from the Penydarren works down the Taf valley to Abercynon. Ironworks at the upper end of the Dulais and Tawe valleys also used the Brecon Forest Tramroad, built in the 1820s as a horse-drawn mineral railway mainly for the carriage of lime. Tramroads are identifiable by the stone setts used to bolt the L-shaped plates to the ground. Of the earlier wooden waggonways, a substantial mid-eighteenth-century section was excavated at Bersham ironworks, and a fragment of similar date has been discovered at Bedlam furnaces.

At Blaenavon there are only partial remains of the ironworks. The original site, founded in 1789, is the best preserved ironworks of its period, but related sites have been cleared entirely. From 1861 when its new blast furnaces entered production, the focus of the Blaenavon Iron and Coal Co. was a new site on the south side of Blaenavon town known as Forgeside, a site that has been redeveloped for light industry. Of an earlier forge at Garnddyrys, erected in 1816 on a bleak upland site below the Blorenge, substantial earthworks remain visible above ground, mainly connected with the water supply. The site has great archaeological potential as an early nineteenth-century forge

Ironmaking

89 Hills Tramroad of 1816, built from Blaenavon north to the Brecknock and Abergavenny Canal. L-shaped iron plates were originally nailed to the stone setts. (Crown copyright: RCAHMW)

that ceased work long before steelmaking was introduced by the company, and where there has been no re-use of the site except by wandering sheep.

The main Blaenavon site retains one of its original blast furnaces, with casting house (the other two were demolished to make way for hot blast stoves in the late nineteenth century), and two furnaces added in around 1810. The base of another furnace, erected in 1860, is all that survives following its demolition and salvage for scrap in 1930. It also has well-preserved ancillary features, including a bank of calcining kilns and a water-balance tower. The latter was built in a modernisation programme initiated in 1839. Its purpose was to raise pig iron from the casting level to the upper level of the works, from where iron was conveyed to the Brecknock and Abergavenny Canal by means of Hills Tramroad, the route of which can be traced almost in its entirety. Opposite the blast furnaces stands a U-shaped block of houses known as Engine Row and Stack Square, built in the period 1789-92 and originally incorporating a truck shop. The proximity of industry and its settlement is the key factor in preserving the compact character of the eighteenth-century ironworks.

The area immediately surrounding the works has been substantially re-landscaped, involving the removal of spoil tips and demolition of company dwellings. Many of these early industrial rows – Chapel Row, Shepherd's Square, Bunker's Row, Upper New Rank – were recorded before their demolition in the 1970s. Evidence of two early shaft mines survives, but most of the early shaft and drift mines have been lost in large-scale landscaping. On the north side of Blaenavon is evidence of ironstone mining and limestone quarrying from the late eighteenth century. Contemporary and earlier evidence of mining coal by means of bell pits, shown in mid-twentieth-century aerial photographs, was lost in the post-war open-cast mining of much of the landscape,

Archaeology and conservation

90 *Above* Furnace 4 at Blaenavon Ironworks, built c.1810, modified c.1880. In 1910 much of its masonry facing was removed to provide building stone for a local parish church. (Crown copyright: RCAHMW)

91 *Right* Blaenavon Ironworks water-balance tower built in 1839, used for lifting and lowering raw materials and finished iron from the two levels of the ironworks. (Crown copyright: RCAHMW)

resulting in significant losses to the archaeological record. On the hills above Blaenavon, ironstone was mined for charcoal-fuelled furnaces in Monmouthshire owned by the Hanbury family of Pontypool, and a geological survey of 1861 claimed that cinder from earlier bloomeries could be found on the open hillsides. Coal was also mined in this area for at least a century before the Blaenavon ironworks was constructed, including for the forges and wire works at Pontypool, 8 miles further south.

Evidence of scouring and patching at Blaenavon is among the best-preserved remains of surface mining in South Wales – comparable to similar recorded landscapes in the minefields of Dowlais, Penydarren and Cyfarthfa Ironworks. Where minerals lay close to the surface, one method of extraction was simply to dig it up, known as patching. At Cyfarthfa in the 1780s the superintendent of the mines was given a telescope by which means he could monitor the performance of the labourers, each of whom was assigned a different patch of ground. But the more prevalent form of ironstone mining was scouring, or hushing, whereby periodic release of water from temporary ponds removed the topsoil to expose the mineral deposits beneath. It was also used to wash through excavated ironstone deposits, separating the dense nodules of ironstone from the clay and shale. The many examples at Blaenavon show the manner in which an outcrop of ironstone was followed progressively, leaving a linear gash across the landscape, with the large deposits of spoil heaped up alongside it. Once exposed, short adits or drifts were dug to exploit the vein below ground.

Scouring had a categorically deleterious effect on the landscape, creating lengths of artificial cliffs, irregular ground surfaces, choked streams and boggy ground. The work was rough, basic outdoor toil not dissimilar to the lot of the agricultural labourer. Except that ironstone was dug and loaded into tramroad wagons in a bleak but preternaturally

92 The landscape of patching and scouring above Blaenavon Ironworks. Deep channels represent the scours, with vast mounds of waste material. (Crown copyright: RCAHMW)

Archaeology and conservation

busy landscape, the atmosphere of which was captured by Svedenstierna in 1802 crossing the hills above Dowlais:

> Higher up, the countryside took on a wild and desert appearance; the so-called roadway was almost unusable for wheeled vehicles; but all around on the slopes of the hills appeared railways, which crossed between the coal mines and several ironworks [i.e. mines] in the distance. About half way between Sirhowy and Merthyr we had a sight which caused us no little amazement. This was a number of mules, laden with coal and ironstone, which were in baskets hung over pack saddles.

People lived close by. At Blaenavon there was a settlement at Pwll Du, close to the major limestone quarries, large enough to have a chapel, and with rows of houses similar to those built close to the ironworks. When it was established, the original owners of the ironworks built a row of small back-to-back dwellings for miners at Bunkers Row (now demolished), of a considerably lower standard than the iron workmen's houses at Stack Square. Elsewhere on the hillsides remain the foundations of houses built singly or in rows of two or three, close to the main workings. These squatter-type dwellings were probably not earlier than the main expansion of mining from 1789. Typically they have generous-sized walled enclosures for pasturing cattle, and probably incorporated garden plots for growing vegetables.

Surface mining belongs to the late eighteenth and early nineteenth century. Later working has left quite different evidence. Ironstone mining gradually declined at Blaenavon, partly because a greater proportion of iron ore was imported, and surface remains of underground workings are minimal. As the nineteenth century progressed,

93 Pwrcas Cottages, a pair of isolated mid-nineteenth-century miners' cottages near Blaenavon. Many of the houses in the district associated with mineral extraction were isolated pairs such as these. (Crown copyright: RCAHMW)

coal gradually increased in importance until it became separated from the iron industry. During the first half of the nineteenth century water-balance lifts were used to raise materials to the surface. The technology was simple and was adapted to other uses – the balance tower at the Blaenavon ironworks has been mentioned above, and the leats, pond and shaft of a similar lift can be seen at Pwll Du limestone quarries. These simple hydraulic lifts had a pulley over a shaft, allowing wagons to be lifted and lowered, using a water-filled tank as a counterweight. The water was released at the bottom, which was only possible in shallow self-draining mines. Dowlais and Blaenavon have good surface remains connected with these features. On the hillside above Blaenavon a dense network of leats channelled water to the holding ponds focused upon three collieries, Hill Pit of 1835, Balance Pit, and New Pit of around 1848. The former pit was later deepened and a winding engine installed. Evidence of both phases are visible – balance ponds and a stack base – as well as foundations of workmen's cottages.

The later technological phase of mining – the steam-powered winding of deep mines – is best represented at Blaenavon by Big Pit, sunk in 1860 by the Blaenavon Iron and Coal Co., near the site of two earlier workings, Coity Pits (to become the upcast for Big Pit) and a drift mine called Dodds Slope. The shaft was deepened in 1880. It originally supplied the company's Forgeside works but production continued long after the ironworks had closed, and the pit was operated by the National Coal Board from 1947–80. Its wooden headgear was replaced in 1921 and in 1953 its steam winding engine was replaced by an electric winding engine. As one of the best-preserved colliery sites in South Wales it signifies the final phase of two centuries of mining activity still visible in the Blaenavon landscape. Limestone quarrying in the region can also be studied chronologically, from the small-scale workings on the north side of the Blorenge to the late nineteenth-century quarry at Cefn-y-lan, each served by its own tramroads.

Mining in Furness has an even longer time-scale, since it reached its peak in the late nineteenth century, when hematite was required for the new furnaces at Barrow and elsewhere. Nab Gill, the most productive of the Eskdale mines in the nineteenth century, had five adits but the surface remains are mostly in the form of spoil heaps and subsidence of underground workings. Tracks reveal how the ore was carried by packhorses down the hillside, until the 1870s when inclined planes were built. Stone Closes, near Stainton in Low Furness, was exploited in the eighteenth century, but the principal evidence here is from the second half of the nineteenth century when two shaft mines were sunk, served by the mineral branch of the Furness Railway, which opened in 1866. Both shafts have associated spoil heaps that were used to infill the shafts after closure. Similarly both had pumping engines that were dismantled, either for scrap or re-use, while the engine houses were also taken down, presumably as a good source of building stone.

Other ore fields were opened up in the nineteenth century. The most successful of several attempts to mine ore from south-west England was at Brendon Hill in Somerset. Large-scale mining, by means of drifts and shafts, began after a mineral lease was signed in 1853 for the supply of ore to the various Ebbw Vale Co. ironworks. Prosperity was short-lived. After the peak years of 1874–78 output declined rapidly, unable to compete with cheaper Spanish ore and at the mercy of the fluctuating fortunes of the Ebbw Vale Co. Brendon Hill had hematite and spathic ores of high quality but they were in thin veins. Transport by means of a railway with inclined plane, then by sea across the Bristol Channel and another transhipment in South Wales was cumbersome, and there were no significant return cargoes – unlike the vessels carrying Spanish ore, which returned carrying coal. Mining at Brendon Hill had ceased by 1883. The centre of mining activity is at the top of an inclined plane, and includes the remains of engine houses, spoil tips and evidence of earlier surface workings. Engine houses such as the surviving building at Burrow Hill Farm, are said to have been built with Cornish expertise and are certainly Cornish in character.

Brendon Hill was also a mining village, albeit a scattered one. From nothing in 1851 there were some sixty cottages housing 250 people by the late 1870s. Most were in stone-built cottages, although there were also 'turf huts' and wooden houses built for railway navvies. There were no public houses – it was a company village run on the abstinence principle – but there were two chapels and an iron mission church, a Miner's Literary Institute, drum bands and a choir. When the mines closed the village died and its stranded inhabitants had no option but to leave. The flourishing village of the 1870s was described by J.L.W. Page in *Exploration of Exmoor* in 1890:

> The gaunt chimneys, the ugly pumping houses, of the deserted iron-mines, do not improve a landscape already rendered sufficiently dreary by the rows of ruinous cottages bordering the roadside. There is in particular a chapel, inscribed 'Beulah', whose blistering walls, boarded windows, and overthrown railings are a sad commentary on this title. A parish doctor told me he could remember the time when over one hundred families of miners occupied the village; now, with the exception of half a dozen cottages let at next to nothing, the place is worse than Goldsmith's village.

Ironically, Beulah is still in use as a chapel, but the remainder of the village has all but gone. Brendon Hill has great archaeological potential, but its most important lesson is the transience of the industrial community.

94 Burrow Farm engine house on the Brendon Hills in Somerset was built in around 1860 for pumping and winding and worked sporadically for less than a quarter of a century. (© author)

SELECT BIBLIOGRAPHY

Addis, John P., 1957, *The Crawshay Dynasty: A study in industrial organisation and development, 1765–1867*. Cardiff: University of Wales Press

Agricola, Georgius, 1556, *De Re Metallica*, translated by H.C and L.H. Hoover. New York: Dover Publications,1950

Allen, G.C., 1929, *The Industrial Development of Birmingham and the Black Country, 1860–1927*. London: George Allen & Unwin Ltd

Angerstein, R.R., 2001, *R.R. Angerstein's Illustrated Travel Diary, 1753–1755: Industry in England and Wales from a Swedish perspective*, translated by Torsten and Peter Berg. London: Science Museum

Ashton, T.S., 1951, *Iron and Steel in the Industrial Revolution*, 2nd edition. Manchester: Manchester University Press

Atkinson, Michael, and Colin Baber, 1987, *The Growth and Decline of the South Wales Iron Industry 1760–1880*. Cardiff: University of Wales Press

Awty, B.G., 1957, 'Charcoal ironmasters of Cheshire and Lancashire, 1600–1785', *Transactions of the Historical Society of Lancashire and Cheshire* 109, 71–124

Awty, B.G., 1981, 'The continental origins of Wealden ironworkers, 1451–1544', *Economic History Review* 34, 524–39

Awty, B.G., and C.B. Phillips, 1979–80, 'The Cumbrian bloomery forge in the seventeenth century and forge equipment in the charcoal iron industry', *Transactions of the Newcomen Society* 51, 25–40

Baber, Colin (ed), 1998, *G.T. Clark: Scholar Ironmaster of the Victorian Age*. Cardiff: University of Wales Press

Barber, Chris, 2002, *Exploring Blaenavon Industrial Landscape*. Abergavenny: Blorenge Books

Barraclough, K.C., 1984, *Steelmaking Before Bessemer, Vol 1: Blister Steel, the Birth of an Industry*. London: The Metals Society

Barraclough, K.C., 1984, *Steelmaking Before Bessemer, Vol 2: Crucible Steel, the Growth of Technology*. London: The Metals Society

Barraclough, K.C., 1990, *Steelmaking: 1850–1900*. London: The Institute of Metals

Bashforth, G.R., 1948, *The Manufacture of Iron and Steel*. London: Chapman & Hall

Beddoes, Thomas, 1791, 'An Account of some appearances attending the conversion of cast into malleable iron, in a letter for Thomas Beddoes MD to Sir Joseph Banks Bart', *Philosophical Transactions of the Royal Society* 81, 173–81

Bennett, Henry, 1864, 'On puddling iron by machinery', *Proceedings of the Institute of Mechanical Engineers*, 298–309

Berg, Maxine, 1994, *The Age of Manufactures 1700–1820: Industry, Innovation and Work in Britain*, 2nd edition. London: Routledge

Berg, Maxine, 1994, 'Technological change in Birmingham and Sheffield in the eighteenth century', in P. Clark and P. Corfield (eds), *Industry and Urbanisation in Eighteenth-Century England*, 20–32. Leicester, Centre for Urban History, University of Leicester

Berg, Maxine and Kristine Bruland (eds), 1998, *Technological Revolutions in Europe: Historical Perspectives*. Cheltenham: Edward Elgar

Birch, Alan, 1967, *The Economic History of the British Iron and Steel Industry 1784–1879*. Manchester: Manchester University Press

Blair, John and Nigel Ramsay (eds), 1991, *English Medieval Industries: Craftsmen, Techniques, Products*. London: Hambledon

Select bibliography

Bowden, Mark (ed), 2000, *Furness Iron: The Physical Remains of the Iron Industry and Related Woodland Industries of Furness and Southern Lakeland*. Swindon: English Heritage

Boyns, T. (ed), 1997, *The Steel Industry Vol 1: The Iron Era: pre 1870*. London: I.B. Taurus

Boyns, T. (ed), 1997, *The Steel Industry Vol 2: The coming of mass steel production 1870–1914*. London: I.B. Taurus

Brooke, E.H., 1944, *Chronology of the Tinplate Works of Great Britain*. Cardiff: William Lewis

Butt, J., 1965–6, 'The Scottish iron and steel industry before the hot-blast', *Journal of the West of Scotland Iron and Steel Institute* 73, 193–220

Byers, Richard L.M., 2004, *Workington Iron and Steel*. Stroud: Tempus

Campbell, R.H., 1961, *Carron Co.* Edinburgh: Oliver & Boyd

Cantrell, John and Gillian Cookson (eds), 2002, *Henry Maudslay and the Pioneers of the Machine Age*. Stroud: Tempus

Carr, J.C. and W. Taplin, 1962, *History of the British Steel Industry*. Oxford: Blackwell

Caswell, B., J. Gaydon and M. Warrender Price, 2002, *Ebbw Vale: The Works 1790–2002*. Ebbw Vale: Works Museum

Chaloner, W., and B.M. Ratcliffe (eds), 1977, *Trade and Transport: Essays in economic history in honour of T.S. Willans*. Manchester: Manchester University Press

Clark, Peter and Penelope Corfield (eds), 1994, *Industry and urbanisation in Eighteenth-Century England*. Leicester, Centre for Urban History, University of Leicester

Cleere, Henry, and David Crossley, 1995, *The Iron Industry of the Weald*, 2nd edition. Cardiff: Merton Priory Press

Cossons, Neil (ed), 1972, *Rees's Manufacturing Industry (1819–20): A selection from The Cyclopaedia; or Universal Dictionary of Arts, Sciences and Literature*, 5 vols. Newton Abbot: David & Charles

Cossons, Neil and Barrie Trinder, 2002, *The Iron Bridge: Symbol of the industrial revolution*, 2nd edition. Chichester: Phillimore

Court, W.H.B., 1938, *The Rise of the Midland Industries, 1600–1838*. London: Oxford University Press

Cox, Nancy, 1990, 'Imagination and innovation of an industrial pioneer: the first Abraham Darby', *Industrial Archaeology Review* 12/1, 127–44

Cranstone, David (ed), 1985, *The Moira Furnace: A Napoleonic Blast Furnace in Leicestershire*. Coalville: NW Leicestershire District Council

Cranstone, David, 1997, *Derwentcote Steel Furnace: An industrial monument in County Durham*. Lancaster: Lancaster University Archaeological Unit

Crossley, David, 1966, 'The management of a sixteenth-century ironworks', *Economic History Review* 19, 273–88

Crossley, David, 1975, 'Cannon-manufacture at Pippingford, Sussex: the excavation of two iron furnaces of c.1717', *Post-Medieval Archaeology* 9, 1–37

Crossley, David (ed), 1980, *Medieval Industry*. York: CBA Research Report 40

Crossley, David, 1990, *Post-Medieval Archaeology in Britain*. Leicester: Leicester University Press

Crossley, David, 1995, 'The supply of charcoal to the blast furnace in Britain', in G. Magnusson (ed), *The Importance of Ironmaking: Technical Innovation and Social Change*, 367–74. Stockholm: Jerkontorets Bergshistoriska Utscott

Crossley, David, 1995, 'The blast furnace at Rockley, South Yorkshire', *Archaeological Journal* 152, 381–421

Crossley, David, and Denis Ashurst, 1968, 'Excavations at Rockley Smithies, a water-powered bloomery of the sixteenth and seventeenth centuries', *Post-Medieval Archaeology* 2, 10–54

Crossley, David and Richard Saville (eds), 1991, *The Fuller Letters 1728–55: Guns, slaves and finance*. Lewes: Sussex Record Society

Davies, A. Stanley, 1940, 'The charcoal iron industry of Powys Land', *Montgomeryshire Collections* 46, 31–66

Davies, R.S.W. and S. Pollard, 1988, 'The iron industry 1750–1850', in C.H. Feinstein and S. Pollard (eds), *Studies in Capital Formation in the United Kingdom 1750–1920*, 73–104. Oxford: Clarendon Press

den Ouden, Alex, 1981, 'The production of wrought iron in finery hearths, part 1: The finery process and its development', *Historical Metallurgy* 15/2, 63–87

Dickinson, H.W. and A. Lee, 1923–4, 'The Rastricks – civil engineers', *Transactions of the Newcomen Society* 4, 48–63

Dinn, James, 1988, 'Dyfi furnace excavations, 1982–87', *Post-Medieval Archaeology* 22, 111–42

Downes, R.L., 1950, 'The Stour Partnership, 1726–36: A note on landed capital in the iron industry', *Economic History Review* 3, 90–96

Ironmaking

Edwards, Ifor, 1957–60, 'The early ironworks of north-west Shropshire', *Transactions of the Shropshire Archaeological Society* 56, 185–202

Edwards, Ifor, 1960, 'The charcoal iron industry of east Denbighshire, 1630–90', *Transactions of the Denbighshire Historical Society* 9, 23–53

Edwards, Ifor, 1961, 'The charcoal iron industry of Denbighshire, c1690–1770', *Transactions of the Denbighshire Historical Society* 10, 49–97

Edwards, Ifor, 1965, 'Iron production in north Wales, the canal era, 1795–1850', *Transactions of the Denbighshire Historical Society* 14, 141–84

Elbaum, Bernard, 1997, 'The steel industry before World War I', in T. Boyns (ed), *The Steel Industry Vol 2: The coming of mass steel production 1870–1914*, 14–42. London: I.B. Taurus

Elsas, Madeleine (ed), 1960, *Iron in the Making: Dowlais Iron Co. Letters 1762–1860*. Cardiff: Glamorgan County Council/GKN

Ely, Judith, 1992, *The Walkers of Rotherham*. Rotherham: Rotherham MBC

Evans, Chris, 1990, 'Gilbert Gilpin: A witness to the South Wales iron industry in its ascendancy', *Morgannwg* 34, 30–8

Evans, Chris, 1992, 'Manufacturing iron in the north-east during the eighteenth century: the case of Bedlington', *Northern History* 28, 178–96

Evans, Chris, 1993, *'The Labyrinth of Flames': Work and Social Conflict in Early Industrial Merthyr Tydfil*. Cardiff: University of Wales Press

Evans, Chris, 1994, 'Iron puddling: The quest for a new technology in eighteenth-century industry', *Llafur* 6/3, 44–57

Evans, Chris, 1994, 'Merthyr Tydfil in the eighteenth century: urban by default?', in P. Clark and P. Corfield (eds), *Industry and Urbanisation in Eighteenth-Century England*, 11–19. Leicester, Centre for Urban History, University of Leicester

Evans, Chris, 1998, 'A skilled workforce during the transition to industrial society: Forgemen in the British iron trade 1500–1850', *Labour History Review* 63, 143–59

Evans, C. and G.G.L. Hayes (eds), 1991, *The Letterbook of Richard Crawshay 1788–1797*. Cardiff: South Wales Record Society

Evans, Chris, and Göran Rydén, 1998, 'Kinship and the transmission of skills: bar iron production in Britain and Sweden, 1500–1860', in M. Berg and K. Bruland (eds), *Technological Revolutions in Europe: Historical Perspectives*, 188–206. Cheltenham: Edward Elgar

Evans, Chris, and Göran Rydén (eds), 2005, *The Industrial Revolution in Iron: The impact of coal technology in nineteenth-century Europe*. Aldershot: Ashgate.

Feinstein, C.H and S. Pollard (eds), 1988, *Studies in Capital Formation in the United Kingdom 1750–1920*. Oxford: Clarendon Press

Fell, Alfred, 1908, *The Early Iron Industry of Furness and District*. Ulverston: Hume Kitchin

Flinn, M.W., 1958, 'The growth of the English iron industry 1660–1760', *Economic History Review* 11, 144–53

Flinn, M.W., 1959, 'The Lloyds in the early English iron industry', *Business History* 2/1, 21–31

Flinn, M.W., 1962, *Men of Iron: The Crowleys in the Early Iron Industry*. Edinburgh: Edinburgh University Press

Gale, W.K.V., 1963–64, 'Wrought iron: a valediction', *Transactions of the Newcomen Society* 36, 1–11

Gale, W.K.V., 1967, *The British Iron & Steel Industry: A Technical History*. Newton Abbot: David & Charles

Gale, W.K.V., 1971, *The Iron and Steel Industry: A Dictionary of Terms*. Newton Abbot: David & Charles

Gale, W.K.V., 1979, *The Black Country Iron Industry: A Technical History*, 2nd edition. London: The Metals Society

Gale, W.K.V., 1992, 'Researching iron and steel: a personal view', *Industrial Archaeology Review* 15/1, 21–35

Gale, W.K.V., 1998, *Ironworking*, 2nd edition. Princes Risborough: Shire

Gale, W.K.V., and Nicholls, C.R., 1979, *The Lilleshall Co. Ltd: a history 1764–1964*. Ashbourne: Moorland Publishing

Gillispie, C.G. (ed), 1959, *A Diderot Pictorial Encyclopaedia of Trades and Industry*. London: Constable

Gould, Shane, 1994, 'Coke ovens at Vobster Breach Colliery', *Industrial Archaeology Review* 17/1, 79–85

Grenter, Stephen, 1992, 'Bersham Ironworks Excavations, 1987–1990: Interim report', *Industrial Archaeology Review* 14/2, 177–92

Grenter, Stephen, 1993, 'A wooden waggonway complex at Bersham Ironworks, Wrexham', *Industrial Archaeology Review* 15/2, 195–207

Griffiths, Samuel, 1873, *Griffiths' Guide to the Iron Trade of Great Britain*. London: the author (facsimile edition David & Charles, 1967)

Gross, Joseph (ed), 2001, *The Diary of Charles Wood of Cyfarthfa Ironworks in Merthyr Tydfil 1766–67*. Cardiff: Merton Priory Press

Select bibliography

Hammersley, G., 1973, 'The charcoal iron industry and its fuel, 1540–1750', *Economic History Review* 26, 593–613

Hancock, Harold and Norman Wilkinson, 1959–60, 'Joshua Gilpin: An American manufacturer in England and Wales, 1795–1801', *Transactions of the Newcomen Society* 32, 15–28

Harris, J.R., 1988, *The British Iron Industry 1700–1850*. Basingstoke: Macmillan

Harris, J.R., 1998, *Industrial Espionage and Technology Transfer: Britain and France in the Eighteenth Century*. Aldershot: Ashgate

Hart, Cyril, 1971, *The Industrial History of Dean, with an introduction to its industrial archaeology*. Newton Abbot: David & Charles

Hawke, G.R., 1970, *Railways and Economic Growth in England and Wales 1840–1870*. Oxford: Clarendon Press

Hay, G.D., and G.P. Stell, 1984, *Bonawe Iron Furnace*. Edinburgh: HMSO

Hay, G.D., and G.P. Stell, 1986, *Monuments of Industry: An Illustrated Historical Record*. Edinburgh: HMSO

Hayman, Richard, 1989, *Working Iron in Merthyr Tydfil*. Merthyr Tydfil: Merthyr Tydfil Heritage Trust

Hayman, Richard, 1997, 'The archaeologist as witness: Matthew Harvey's Glebeland Works, Walsall', *Industrial Archaeology Review* 19, 61–74

Hayman, Richard, 2004, 'The Cranage brothers and eighteenth-century forge technology', *Historical Metallurgy* 38/2, 113–20

Hayman, Richard, and Wendy Horton, 2003, *Ironbridge: History & Guide*, revised edition. Stroud: Tempus

Hayman, Richard, Wendy Horton and Shelley White, 1999, *Archaeology and Conservation in Ironbridge*. York: CBA Research Report 123

Hey, David, 1972, *The Rural Metalworkers of the Sheffield Region: a study of rural industry before the industrial revolution*. Leicester: Leicester University Press

Hey, David, 1991, *The Fiery Blades of Hallamshire: Sheffield and its neighbourhood, 1660–1740*. Leicester: Leicester University Press

Hopkinson, G.G., 1961, 'The charcoal iron industry of the Sheffield region', *Transactions of the Hunter Archaeological Society* 8, 122–51

Hudson, Pat (ed), 1989, *Regions and Industries: A perspective on the industrial revolution in Britain*. Cambridge: Cambridge University Press

Hughes, Stephen, 1990, *The Archaeology of an Early Railway System: The Brecon Forest Tramroads*. Aberystwyth: RCAHM Wales

Hyde, Charles, 1977, *Technological Change in the British Iron Industry 1700–1870*. Princeton: Princeton University Press

Ince, Laurence, 1991, *The Knight Family and the British Iron Industry*. Birmingham: Ferric Publications

Ince, Laurence, 1992, 'The Boulton and Watt engine and the British iron industry', *Wilkinson Studies* 2, 81–89

Ince, Laurence, 1993, *The South Wales Iron Industry, 1750–1885*. Birmingham: Ferric Publications

Ince, Laurence, 2001, *Neath Abbey and the Industrial Revolution*. Stroud: Tempus

Jenkins, Paul, 1995, *'Twenty by Fourteen': A History of the South Wales Tinplate Industry 1700–1961*. Llandysul: Gomer

Johnson, B.L.C., 1951, 'The charcoal iron industry in the early eighteenth century', *Geographical Journal* 117, 167–77

Johnson, B.L.C., 1952, 'The Foley partnerships: the iron industry at the end of the charcoal era', *Economic History Review* 4, 322–40

Johnson, B.L.C., 1953, 'New light on the iron industry of the Forest of Dean', *Transactions of the Bristol & Gloucestershire Archaeological Society* 72, 129–43

Johnson, B.L.C., 1960, 'The Midland iron industry in the early eighteenth century: The background to the first successful use of coke in iron smelting', *Business History* 2/2, 67–74

Johnson, M.P. and P. Worrall, 1983, *Top Forge, Wortley*, 2nd edition. Sheffield: Sheffield Trades Historical Society

Jones, Edgar, 1987, *A History of GKN, volume 1: Innovation and Enterprise, 1759–1918*. London: Macmillan

Joyce, Patrick, 1980, *Work, Society and Politics: The Culture of the Factory in Later Victorian England*. London: Harvester Press

Joyce, Patrick (ed), 1987, *The Historical Meanings of Work*. Cambridge: Cambridge University Press

Joynson, F. (ed), 1892, *The Iron and Steel Maker*. London: Ward, Lock, Bowden & Co

Ironmaking

Kanefsky, John, and John Robey, 1980, 'Steam engines in eighteenth-century Britain', *Technology and Culture* 21/2, 161–87

King, P.W., 1996, 'Early statistics for the iron industry: a vindication', *Historical Metallurgy* 30/1, 23–45

King, P.W., 1997, 'The development of the iron industry in South Staffordshire in the seventeenth century: history and myth', *Transactions of the Staffordshire Archaeological and Historical Society* 38, 59–76

Knight, Jeremy, 1992, *Blaenavon Ironworks*. Cardiff: Cadw

Kohn, Frederick, 1869, *Iron and Steel Manufacture*. London: William Mackenzie

Labouchere, Rachel, 1988, *Abiah Darby, 1716–1793, of Coalbrookdale: Wife of Abraham Darby II*. York: William Sessions

Lane, Michael R., 1993, *The Story of the Britannia Ironworks: William Marshall, Sons & Co, Gainsborough*. London: Quiller Press

Le Guillou, M., 1972, 'The South Staffordshire iron and steel industry and the growth of foreign competition (1850–1914), Part 1', *West Midlands Studies* 5, 16–23

Le Guillou, M., 1973, 'The South Staffordshire iron and steel industry and the growth of foreign competition (1850–1914), Part 2', *West Midlands Studies* 6, 41–45

Lead, Peter, 1977, 'The North Staffordshire iron industry 1600–1800', *Historical Metallurgy* 11/1, 1–14

Lewis, J.H., 1984, 'The charcoal-fired blast furnaces of Scotland: a review', *Proceedings of the Society of Antiquaries of Scotland* 114, 433–79

Lindsay, J.M, 1977, 'The iron industry in the Highlands: charcoal blast furnaces', *Scottish Historical Review* 56, 49–63

Linsley, S.M., and R. Hetherington, 1978, 'A seventeenth-century blast furnace at Allensford, Northumberland', *Historical Metallurgy* 12, 1–11

Littlewood, Kevin, 1999, 'Rhymney's Egyptian Revival: Images and interpretations of the Bute Ironworks, Glamorganshire, 1824–44', *National Library of Wales Journal* 31/1, 11–39

Lloyd, Humphrey, 1975, *The Quaker Lloyds in the Industrial Revolution*. London: Hutchinson

Lord, Peter, 1998, *The Visual Culture of Wales: Industrial Society*. Cardiff: University of Wales Press

Lowe, Jeremy, 1985, *Welsh Industrial Workers' Housing*, 2nd edition. Cardiff: National Museum of Wales

Lowe, Jeremy, and Martin Lawlor, 1980, 'Landscapes of the iron industry at Blaenavon, Gwent', *Landscape History* 2, 74–82

Macleod, Christine, 1988, *Inventing the Industrial Revolution: The English Patent System 1660–1800*. Cambridge: Cambridge University Press

Macleod, Christine, 1998, 'James Watt, heroic invention and the idea of the industrial revolution', in M. Berg and K. Bruland (eds), *Technological Revolutions in Europe: Historical Perspectives*, 96–116. Cheltenham: Edward Elgar

Magnusson, G. (ed), 1995, *The Importance of Ironmaking: Technical Innovation and Social Change*. Stockholm: Jerkontorets Bergshistoriska Utscott

Mathias, Peter, 1979, *The Transformation of England: Essays in the Economic and Social History of England in the Eighteenth Century*. London: Methuen

Minchinton, W.E., 1957, *The British Tinplate Industry: A History*. Oxford: Clarendon Press

Mitchell, B.R., and P. Deane, 1962, *Abstract of British Historical Statistics*. Cambridge: Cambridge University Press

Morton, G.R., 1973, 'The wrought-iron trade of the West Midlands', *West Midlands Studies* 6, 5–18

Morton, G.R. and N. Mutton, 1967, 'The transition to Cort's puddling process', *Journal of the Iron and Steel Institute* 205, 722–28

Mott, R.A., 1983, *Henry Cort: The Great Finer*. London: The Metals Society

Mushet, David, 1840, *Papers on Iron and Steel, Practical and Experimental*. London: John Weale

Muter, W.G., 1979, *The Buildings of an Industrial Community: Ironbridge and Coalbrookdale*. Chichester: Phillimore

Mutton, Norman, 1965–68, 'The forges at Eardington and Hampton Loade', *Transactions of the Shropshire Archaeological Society* 58, 235–343

Mutton, Norman, 1969, *An Engineer at work in the West Midlands: The Diary of John Urpeth Rastrick for 1820*. Journal of West Midlands Regional Studies Special Publication No 1

Mutton, Norman, 1976, 'The Marked Bar Association: Price regulation in the Black Country wrought-iron trade', *West Midlands Studies* 9, 2–8

Newman, John, 1995, *The Buildings of Wales: Glamorgan*. London: Penguin

Newman, John, 2000, *The Buildings of Wales: Gwent/Monmouthshire*. London: Penguin

Newman, Richard, David Cranstone and Christine Howard-Davies, 2001, *The Historical Archaeology of Britain c1540–1900*. Stroud: Sutton Publishing

Select bibliography

Paar, H.W., and D.G. Tucker, 1975, 'The old wireworks and ironworks of the Angidy valley at Tintern, Gwent', *Historical Metallurgy* 9, 1–14

Page, Robert, 1979, 'Richard and Edward Knight: Ironmasters of Bringewood and Wolverley', *Transactions of the Woolhope Naturalists Field Club* 43/1, 7–17

Payne, Peter, 1979, *Colvilles and the Scottish Steel Industry*. Oxford: Clarendon Press

Peatman, Janet, 1989, 'The Abbeydale industrial hamlet: history and restoration', *Industrial Archaeology Review* 11/2, 141–54

Penfold, Alastair (ed), 1980, *Thomas Telford: Engineer*. London: Thomas Telford Ltd

Percy, John, 1864, *Metallurgy, Section 1: Iron and Steel*. London: John Murray

Phillips, C.B., 1977, 'The Cumbrian iron industry in the seventeenth century', in W.H. Chaloner & B.M. Ratcliffe (eds), *Trade and Transport: Essays in economic history in honour of T.S. Willans*, 1–34. Manchester: Manchester University Press

Pickin, J., 1982, 'Excavations at Abbey Tintern Furnace', *Historical Metallurgy* 16, 1–21

Plot, Robert, 1686, *The Natural History of Staffordshire*. Oxford, printed at the theatre (facsimile edition E.J. Moreton, 1973)

Pollard, Sidney, 1968, *The Genesis of Modern Management: A Study of the Industrial Revolution in Great Britain*. Harmondsworth: Penguin

Porter, J.H., 1969, 'Management, competition and industrial relations: The Midlands manufactured iron trade 1873–1914', *Business History* 11/1, 37–47

RCAHM Wales, 2003, *The Archaeology of the Welsh Uplands*. Aberystwyth: RCAHMW

Raistrick, Arthur, 1953, *Dynasty of Ironfounders*. London: Longmans, Green & Co

Raistrick, Arthur, 1968, *Quakers in Science and Industry: Being an Account of the Quaker Contribution to Science and Industry during the Seventeenth and Eighteenth Centuries*, 2nd edition. Newton Abbot: David & Charles

Raistrick, A. and E. Allen, 1939, 'The South Yorkshire ironmasters', *Economic History Review* 1st series 9, 168–85

Rathbone, Hannah Mary, 1852, *Letters of Richard Reynolds, with a memoir of his life*. London: Charles Gilpin

Riden, Philip, 1977, 'The output of the British iron industry before 1870', *Economic History Review* 30, 442–59

Riden, Philip, 1988, 'The ironworks at Alderwasley and Morley Park', *Derbyshire Archaeological Journal* 108, 77–109

Riden, Philip, 1989, 'The ironworks at Alderwasley and Morley Park: a postscript', *Derbyshire Archaeological Journal* 109, 175–9

Riden, Philip, 1990, *The Butterley Co., 1790–1830*. Chesterfield: Derbyshire Record Society

Riden, Philip, 1991, 'The charcoal iron industry in the East Midlands, 1580–1780', *Derbyshire Archaeological Journal* 111, 64–84

Riden, Philip, 1992, *John Bedford and the Ironworks at Cefn Cribwr*. Cardiff: the author

Riden, Philip, 1992, 'Early ironworks in the lower Taff valley', *Morgannwg* 36, 69–83

Riden, Philip, 1993, *A Gazetteer of Charcoal-fired Furnaces in Great Britain in use since 1660*, 2nd edition. Cardiff: Merton Priory Press

Riden, Philip, 1994, 'The final phase of charcoal iron-smelting in Britain, 1660–1800', *Historical Metallurgy* 28/1, 14–26

Riden, Philip, 1995, 'The blast furnace in Great Britain: survival and conservation', in G. Magnusson (ed), 319–26

Riden, Philip and Owen, John, 1995, *British Blast Furnace Statistics 1790–1980*. Cardiff: Merton Priory Press

Rowlands, Marie B., 1975, *Masters and Men in the West Midland Metalware Trades before the Industrial Revolution*. Manchester: Manchester University Press

Rowlands, Marie B., 1989, 'Continuity and change in an industrialising society: the case of the West Midlands industries', in P. Hudson (ed), *Regions and Industries: a perspective on the industrial revolution*, 103–31. Cambridge: Cambridge University Press

Scarfe, Norman (ed), 1995, *Innocent Espionage: The La Rochefoucauld Brothers' Tour of England in 1785*. Woodbridge: Boydell Press

Schafer, R.G., 1971, 'Genesis and structure of the Foley Ironworks in Partnership', *Business History* 13/1, 19–38

Schubert, H.R., 1957, *History of the British Iron and Steel Industry from c450 BC to AD 1775*. London: Routledge & Kegan Paul

Scrivenor, Harry, 1854, *History of the Iron Trade*, 2nd edition. London: Longman, Brown, Green and Longmans

Sellick, Roger, 1970, *The West Somerset Mineral Railway and the Story of the Brendon Hills Iron Mines*, 2nd edition. Newton Abbot: David & Charles

Sim, David, and Isabel Ridge, 2002, *Iron for the Eagles: The Iron Industry of Roman Britain*. Stroud: Tempus

Ironmaking

Smiles, Samuel, 1863, *Industrial Biography: Iron Workers and Tool Makers*. London: John Murray

Smith, W.A., 1970/71, 'The contribution of the Gibbons family to technical development in the iron and coal industries', *West Midlands Studies* 4, 46–54

Smith, W.A., 1978, 'Combinations of West Midlands ironmasters during the industrial revolution', *West Midlands Studies* 11, 1–10

Stones, Frank, 1977, *The British Ferrous Wire Industry 1882–1962*. Sheffield: J.W. Northend

Straker, Ernest, 1931, *Wealden Iron*. London: Bell & Sons

Svedenstierna, Eric, 1973, *Svedenstierna's Tour of Great Britain 1802–3: The Travel Diary of an Industrial Spy*, translated by E.L. Dellow. Newton Abbot: David & Charles

Thomas, Brinley, 1986, 'Was there an energy crisis in Great Britain in the seventeenth century?', *Explorations in Economic History* 23, 124–52

Thompson, E.P., 1991, *Customs in Common*. London: Merlin

Timmins, Samuel (ed), 1866, *The Resources, Products and Industrial History of Birmingham and the Midland Hardware District*. London: Robert Hardwicke (facsimile edition Frank Cass, 1967)

Treadwell, J.M., 1974, 'William Wood and the Co. of Ironmasters of Great Britain', *Business History* 16/2, 97–112

Trinder, Barrie (ed), 1979, *Coalbrookdale 1801: a contemporary description*. Ironbridge: IGMT

Trinder, Barrie, 1982, *The Making of the Industrial Landscape*. London: Dent

Trinder, Barrie (ed), 1988, *'The Most Extraordinary District in the World': Ironbridge and Coalbrookdale*, 2nd edition. Chichester: Phillimore

Trinder, Barrie, 1996, *The Industrial Archaeology of Shropshire*. Chichester: Phillimore

Trinder, Barrie, 2000, *The Industrial Revolution in Shropshire*, 3rd edition. Chichester: Phillimore

Truran, Samuel, 1855, *The Iron Manufacture of Great Britain*. London: E. & F. Spon

Tucker, Gordon and Peter Wakelin, 1981, 'Metallurgy in the Wye valley and South Wales in the late eighteenth century: new information about Redbrook, Tintern, Pontypool and Melingriffith', *Historical Metallurgy* 15, 94–100

Tweedale, Geoffrey, 1995, *Steel City: Entrepreneurship, Strategy & Technology in Sheffield 1743–1993*. Oxford: Clarendon Press

Tylecote, R.F., 1970, *A History of Metallurgy*. London: The Metals Society

Tylecote, R.F., 1986, *The Prehistory of Metallurgy in the British Isles*. London: Institute of Metals

van Laun, John, 1979, 'Seventeenth-century ironmaking in south-west Herefordshire', *Historical Metallurgy* 13, 55–68

Warren, Kenneth, 1990, *Consett Iron 1840–1980: A Study in Industrial Location*. Oxford: Clarendon Press

Watters, Brian, 1998, *Where Iron Runs Like Water: A New History of the Carron Ironworks, 1759–1982*. Edinburgh: John Donald

Wilson, Anne, 1988, 'The excavation of Clydach Ironworks', *Industrial Archaeology Review* 11/1, 16–36

INDEX

Adam, Robert 93
Addenbrooke, John *see* John Homfray
Agricola, Georgius 22, *2, 3, 6*
Angerstein, Reinhold 32-33, 43
Attwick, William 46
Avenant, Richard 21, 28
Bacon, Anthony 48, 100
Bailey Brothers, iron merchants 80
Bailey family, ironmasters
  Crawshay 94
    Joseph 100
    Baird & Co. 81
Baldwin, J.P & W. 83-84
Barrow Hematite Steel Co. 122
Barry, Sir Charles 114
Beale, Samuel & Co. 81
Bedford, John 107
Bedson, Charles 72
Bell, Isaac Lowthian 69
Bennett, Henry 118
Bessemer, Sir Henry 10, 116
bloomeries
  Ashwicken 14
  Bardown 15
  Beauport Park 15
  Byrkeknott1 18, 105
  Chingley 17-18
  Constantine's Cave 13
  Holbeanwood 15
  Lyveden 16
  Muncaster Head 18
  Ramsbury 15-16
  Rockley Smithies 18
  Rudh' an Dunain 13
  Timberholme 20
  Tudeley 16, 18
  Waltham Abbey 16
  West Runton 16
Bolckow, Vaughan & Co. 69, 80, 121
Botfield family
  Beriah III 121
  Thomas 45, 95, 121

  William 57, 121
Boulton & Watt 44, 52
Bradley & Co. 79, 80
brass manufacture 31, 35
Brecon Forest Tramroad 143
Brendon Hill 148-50
brick and tile manufacture 94, 120-21
British Iron Co. 64-65
British Iron Trade Association 57
bronze manufacture 11
Brown, John & Co. 81, 82
Brown, Lancelot 'Capability' 88
Brownrigg, William 43
Brunel, Isambard Kingdom 81, 102
Brunton, William 91
Burton, Decimus 65
Bute, 2nd Marquess of 65-66
Caesar, Julius 13, 14
calcining 13, 14, 23, 74, 127, 131, 141, *3, 44, 88*
Cammell & Co. 82, 117, 119
canals 61-62
Carnegie, Andrew 122
charcoal burning 131, *79*
Cheney, Robert 121
Clark, G.T. 102-03
Cliffe, Charles 7
coal mining 40, 94, 121, 127, 144-46, 148
Coalbrookdale 7, 135-36, *16*
  Literary and Scientific Institution 114, 135, *67*
  see also iron and steel works
Coalbrookdale Co. 52, 76, 93, 114;
  see also iron and steel works
Cockshutt, James 48
coking 35, 73-74, 141, *42*
Condie, John 71
Consett Iron Co. 80, 121
Cordell, Alexander 126
Cort, Henry 8, 10, 46-48

Cotton family 21
Cowper, E.A. 69
Cranage
  Thomas and George, patentees 43, 47
  Thomas, hammerman, 47
Crane, George 64
Crawshay family 58
  Francis 83
  Richard 48-50, 55, 59, 62, 94, 97, 100, 103, 104, 108
  Robert Thompson 120
  William I 48, 100
  William II 83, 100, 121, 141, *58*
Crowley
  Ambrose 32, 97
  John 7
Cyfarthfa Castle 100, 142-43, *59*
Danks, Samuel 118
Darby family 98, 99
  Abraham I 10, 34-36, 94, 133-34, 136
  Abraham II 39-40, 98
  Abraham III 41, 95, 99, 134
  Abraham IV 99
  Alfred 99
Darlington Iron Co. 69
Dawson, Joseph 58
Dixon, William 81
Downton Castle 101-02
Dudley, Dud 35
Dudley, Earl of 71-72
Dudley, Robert, Earl of Leicester 21
Ebbw Vale Co. 65, 148
Edwards, William 109
Fairbairn, William 81
Fell, John 32
Ferriday, William 95
Field, Joshua 92
Filarete, Antonio Averlino 19
Foley family 28
  Paul 21, 28, 31
  Philip 28, 31

Ironmaking

Ford, Edmund, 94
Ford, Richard 39, 136
Forman, William 65-66
Foster, James 84
Gale, Keith 125
George, Watkin 53
Gibbons family 45, 58
　John, iron merchant 31
　John, ironmaster 95, 108
Gilbertus, smith, 17
Gilpin family, agents 95
　Gilbert 53, 57, 59, 95, 99
Gilpin, Joshua, paper manufacturer 55-56, 76, 88
Gjers, John, *44*
Glasgow Iron Co. 81
Glynn, Joseph 91
Goldney, Thomas I and II 98
Goodrich, Simon 76
Granville, Earl 66
Great Exhibition 80, 93
Grenfell, G.N. 107
Guest family
　John 62, 95
　Sir Josiah John 95, 102, 121
Hall, Joseph 70
Hallen family 30-31
　Samuel 45
Hanbury, John 23-24, 82
Handyside, Andrew 93
Hanney & Sons 81
Harvey, Thomas, ironmaster 31
Harvey & Co., founders 91
Herbert, William, Earl of Pembroke 21, 94
Hill family 62
Hoare, Richard Colt *27*
Holland family *see* Hallen family
Homfray family 97-98
　Francis III 55, 98
　Jeremiah 55, 62, 97, 98
　John 98
　Samuel 55, 62, 97, 98
Hood, Thomas and William 60
Houliere, Marchant de la 43
houses
　ironmaster 98, 99-103, *57, 59*
　workmen 109-14, 127, 135, 136, 142-43, 144, 147, *62, 63, 64, 65, 74, 82, 93*
Humfrey, William 29-30
Huntsman, Benjamin 82
Ibbetson, Julius Caesar *1*
Inchtuthil hoard 15
iron, manufactured 15, 17, 18, 31, 32-33, 86
　aqueducts 41, 90, 92, *21, 50*
　armour 17
　armour plates 81
　art castings 93, *53*
　bridges 41, 88-90, *20, 51, 57*

cooking pots 35
in architecture 15, 17, 41, 90-91, 92-93
Iron Bridge (Shropshire) 41, *20*
locomotive manufacture 80-81
nail trade 29, 32
ordnance 24, 41, 48
pan-making 30-31
rails 79-80, 119
ship plates 81, 92, 119
weapons 13-14, 15
wire 85
see also steam engines
Iron and Steel Institute 118, 119
iron and steel works
　Abercarn 38, 128
　Aberdare 52
　Abernant 91
　Abersychan 64-65, 80, 99
　Allensford 131
　Atlas (Glasgow) *40*
　Atlas (Sheffield) 81, 82, 116, 118
　Backbarrow 20, 37, 131
　Banwen 64, 138
　Barrow 122, *38, 69, 70, 71*
　Beaufort 52, 96
　Bedlam 40, 52, 53, 86, 94, 121, 136, 141, 143, *83*
　Bedlington 79
　Bersham 39, 40, 86, 95, 137-38, 141, 143
　Blaenavon 8, 52, 53, 62, 80, 105-06, 109, 122, 127, 141, 143, 144, *27, 88, 90, 91*
　Blaina 64, 80
　Blists Hill 127, 136, 137, 139, 141, *36*
　Blochairn 81
　Bloomfield *45*
　Bonawe 38, 111, 128, 129, *17, 77, 78*
　Bowling 81
　Bradley 44, 55, 107
　Brightside 82
　Bringewood 28, 100, 102
　Britannia (Gainsborough) 92
　Britannia (Middlesbrough) 69
　Broadwaters 56, 84, 97
　Brookland 19
　Brymbo 44, 137
　Bryncoch 39
　Burblethwaite 20
　Bute 65-66
　Butterley 80-90-91, 111
　Buxted 24
　Caerleon 38
　Caerphilly 128
　Calcutts *19*
　Cannop 23, 110
　Capponfield 80
　Carmarthen 128

Carnforth *72*
Carron 40, 43, 47, 86, 93
Castle 118
Cefn Cribwr 107, 138
Cefn Cwsk 141
Charlcotte 28, 128
Chillington 80, 118
Cincinnati Railway Ironworks (USA) 118
Clarence 69
Cleobury Dale 45
Cleobury Mortimer 21
Clifton 39, 81
Clydach 52, 53, 109, 129, 139, 143, *28*
Coalbrookdale 31, 34-35, 36-37, 43, 44, 52, 53, 76, 88, 91, 95, 98, 99, 109, 133-35, 137, *14, 16, 22, 81*
Coalbrookvale 64
Coats 81
Congreave 72
Coniston 20
Consett 69, 80, 119
Cookley 28, 40, 84
Corbyns Hall 95
Cradley 45
Craleckan 38
Cunsey 20, 131
Cwm Celyn 64, 76
Cwmavon 80
Cyclops (Sheffield) 82, 116, 118, 119
Cyfarthfa 40, 41, 43, 48-49, 50, 53-54, 79, 94, 97-98, 100, 108, 120, 121, 141-42, *23, 25, 26*
Dalmellington 141
Dark Lane 84
Darlaston 85, 118, *34*
Derwent 122
Dol-gun 128
Dolobran 27, 107
Donnington Wood 44, 66
Dowlais 40, 50, 59, 60, 77, 79, 80, 95, 102, 109, 117, 118, 121, 129, *31, 85*
Drumpeller 81
Duddon 37, 128, 129-31
Dulais 54
Dyfi 128-29, *18, 76*
Eardington 84
East Moors 102, 121
Ebbw Vale 52, 53, 70, 76, 77, 79, 98, 99, *39*
Elsecar *41*
Etruria 66, *35*
Flaxley 28
Fontley 46
Force 20
Forgeside (Blaenavon) 143
Gadlys 64, 139, 141
Garnddyrys 53, 126, 143-44
Glangrwyne 54

Index

Glen Kinglass 38
Gloucester (Sussex) 23
Golynos 64
Gothersley 55
Govan 81
Greensforge 27
Gunns Mill 27
Hacket 20
Hampton Loade 84
Heath 27
Hirwaun 40, 143
Holytown 81
Horsehay 39, 44, 56, 60, 105, 109, *30*
Horseley 90
Hyde 47, 98
Invergarry 38
Ketley 39, 44, 47, 55-56, 76, 120-21
King's Hill 118
Kirkstall 81
LNWR Crewe 72, 117
Landore 117, 119
Lawton 21
Leighton (Furness) 23, 24, 37, 131
Leighton (Shropshire) 94
Level 45, 95, 108
Lightmoor 40, 55, 98
Lizard 21
Llynfi 64, 139-41, *87*
Lodge 66, *37*
Low Mill 42, 43
Low Moor 58, 81
Lowwood 131
Lydbrook 21, 22, 23, 110
Lydney 21
Lye 45
Madeley Wood *see* Bedlam
Manchester Wire Works 72
Mathrafal 27
Mayfield 19
Meir Heath 28
Melincourt 128
Merryston 81
Mitton 40, 44, 56
Moira 138, *84*
Monway 85
Moreton Corbet 28, 95
Moss Bay 122
Motherwell 81
Nantyglo 52, 53, 62, 70, 79, 94, 143
Neath Abbey 52, 91, 141
Netherton 76, 139
New Willey 40, 44
Newbridge 20
Newland 76, 131
Nibthwaite 27-28, 131
Oakwood 54
Old Park (Shropshire) 44, 45, 55, 56, 57, 59, 60, 61, 78, 84-85, 95, 105, 107, 108, 109, 121

Old Park (Staffordshire) 81, 117
Onllwyn 64
Ormesby 69
Parkend 21, 109-10
Parkgate 81, 82
Parkhead 81, 92
Penny Bridge 131
Pentwyn 64, 99
Penydarren 49, 50, 52-53, 55, 79, 91, 95, 98, 121, 143, *24*
Perran 91-92
Phoenix (Glasgow) 81
Pippingford 24
Pitchford 45
Plas Kynaston 92
Plymouth 40, 79, 121, 127
Pontnewydd 80
Pontypool 38, 54, 82, 146
Priorslee 66, 139
Rhymney 52, 66, 79, 80
Rochsolloch 81
Rockley 127, *75*
Rotherhithe 48
Round Foundry 91
Round Oak 71-72, 81, 118, 125, *73*
Russell's Hall 76
St Rollox 81
Sample 117
Saracen (Glasgow) 93
Scarlets 24
Seend 141
Shifnal 21
Shut End 47
Sirhowy 52, 95, 99
Snedshill 44
Snedshill Forge 66, 85
Soho 91, 95
Soudley 21
Spark Bridge 20
Stirchley 78, 85, 105
Stony Hazel 132
Stourton 27, 28
Stumbletts 24
Sutton 39
Swindon (Staffordshire) 27
Taff Vale 72
Tees Side 69
Tern 31
Tibberton 45
Tintern 28, 29-30, 127
Tividale 80
Tondu 141
Tredegar 52, 79, 98
Trellech 127
Trostre 107
Union *see* Rhymney
Upton 45, 92, 107
Vale Royal 21
Varteg 120
Victoria (Derby) 118

Victoria (Monmouthshire) 64, 99
Vulcan (Glasgow) 92
Walker 69
Wednesbury 42, 43, 45
Wenallt 64
West Bromwich 43
West Cumberland 122
Westbury 141
Whitchurch 21
Whitebrook 30
Whitecliff 138
Whitehill 39
Whittington (Derbyshire) 118
Whittington (Staffordshire) 27
Wilden 28, 44, 56, 83-84, *85*
Willey 39
Witton Park 69
Wolverley 28, 40, 100
Wombridge 118
Woolbridge 19
Workington 122
Wortley 81, 132-33, *80*
Wrens Nest 44
Ynyscedwyn 64
Ynysfach 50, 139-41, 142, *86*
Ystalyfera 64, 82
see also bloomeries, tinplate works
iron mining 12-13, 14, 18, 105-06, 121, 125-25, 131, 146-47, 148-50, *43, 61, 73, 92, 94*
Ironbridge Gorge Museums 134
ironmaking
   air furnaces *see* reverberatory furnaces
   blast furnace 22-24, 76-77, *8*
   bloomery 11-12
   casting 87-88, *47, 48, 49*
   cupola furnaces 86-88, *46*
   finery 25, *9, 10*
   hot blast 66, 69
   mechanical puddling 118
   Osmond iron 30
   pig boiling 70
   puddling 47, 49, 55-56, 77-78
   refineries *see* running-out fires
   reverberatory furnaces 35-36, 42, 86, *15*
   rolling mills 46-47, 58-59, 72, 78-79, *41*
   running-out fires 49, 77
   slitting mills 29, *13*
   snapper furnace 134
   stamping and potting 42-44
   steam hammers 71-72, *40*
   wire drawing 29-30, 72
   see also steam engines, water power
Ironworks in Partnership 28, 95
Jars, Gabriel 108
Jeans, J.S. 119
Jefferson, Thomas 41

Ironmaking

Jellicoe
  Adam 46
  Samuel 46, 48
Jesson, Richard *see* Wright & Jesson
Jessop, William 90, 91
Jones, Thomas 47
Kendall, Edward 96, 103
Keysar, Barnes 30
Knight family 100-02
  Edward 28, 40, 57, 100
  John 84
  Richard 28, 95, 100
  Richard Payne 100-02
Knott, George 111
Krupp, Alfred 121, 122
labour force
  ale allowance 108
  boys 106
  immigrant 20, 104
  medical welfare 109
  religion 114
  truck system 113-14
  women 105-6
  see also houses, schools
Laird, John 81
Lavender family 107
Levett, William 24
Lewis, William 52
Lilleshall Co. 66
limestone quarries 148
Lloyd, Charles 107
Llyn Cerrig Bach hoard 13
Llyn Fawr hoard 13
Lugar, Robert 100, 143
Macculloch, Dr James 65-66
Macfarlane & Co. 93
Malkin, Benjamin 112
Martin, Pierre and Emil 117
Maudslay, Henry 92
McDowall Stevens & Co. 93
Menelaus, William 80, 121
Mersey Iron & Steel Co. 81
Merthyr Tydfil 7, 111-14
Mossend Iron Co. 81
Munby, A.J. 105-6
Museum of Welsh Life 143
Mushet, David 138
Napier, Robert 81, 92
Nash, John 93
Nasmyth, James 71
Neilson, James 10, 66
Nettlefold & Chamberlain 118
Newcomen, Thomas 36
Onions, Peter 47

Outram, Benjamin 111
Page, J.L.W. 149-50
Pamplin, William *23*
Patent Shaft & Axletree Co. 81
Pentwyn & Golynos Co. 80
Percy, Dr John 8, 10, 78, 104, 188, 120
Plot, Dr Robert 30
Postlethwaite, Malachy 46
Price, Joseph Tregelles 91
Pritchard, Thomas Farnolls 41
public houses 114
Ramsbottom, John 72
Rastrick, John Urpeth 69
Rawlinson, William 37-38
Ray, John 23
Rennie, John 91
Reynolds family 98
  Joseph 47, 99
  Richard 43, 98-99, *56*
  William 7, 47, 55, 97, 99, *55*
Rhymney Iron Co. 80, 83; *see also* iron and steel works
Robertson, James & Co 61
Rochefoucauld, Alexandre and François de la 76, 114
Roebuck, Dr John 43
Rogers, Samuel 70
Rosedale, calcining kilns 141
Royal Mint 29-30
Savery, Thomas 36
schools, iron-company 114
Schutz, Christopher 30
Scott, Giles Gilbert 93
Scrivenor, Harry 8
Severn, River 32, 61, 63
Sheward, James 61
Siemens family
  Frederick 117
  Karl 118
  William 117
Smeaton, John 40, 86
Smiles, Samuel 10
Smith, George Kenrick 120
Smith-Casson, Edward 118
Smitheman, John 94
Smyth, Warington *36*, *43*
Spencer family 21, 32
Spooner, Abraham 27, 28, 29
Staffordshire Works 28
steam engines 36-37, 39-40, 44, 52-53, 69-70, 79, 91-92, *38*, *39*, *52*, *71*, *72*
steel

acid Bessemer 116-17, *68*
basic Bessemer 122
blister 30
crucible 82
in medieval Britain 17
in Roman Britain 15
open-hearth 117
shear 30
Strabo 13
Svedenstierna, Eric 86-87, 108
Taitt, William 60
Talbot, George 21
Talygarn, country house 102-03
Telford, Thomas 88, 90
Thames Iron Co. 81
Thomas of Leighton 17
Thomas, Sidney Gilchrist 122
Thompson, Robert 109
tinplate manufacture 82-83
tinplate works
  Treforest 83, 141, 143
  Vernon 82
  see also iron and steel works
trades unions 119-20
tramroads 62-63, 143, 144, *63*, *89*
Tredegar, Lord 98
Trevithick, Richard 91
Tubbe, John 16
Turner, J.M.W. 142
Vaughan, John *see* Bolckow, Vaughan & Co.
Vivares, Francois *37*
water power 17-18, 19, 27, 53-54, 79, 142
Watt, James, 44, 91; *see also* Boulton & Watt
Weale, James 8
Wheeler, John 21, 28, 31, 95
While, Charles 72
Wilkinson
  Isaac 40, 86, 95, 137
  John 40-41, 44, 55, 86, 96, 97, 103, 114, 137, *54*
  William 86, 137
Williams, Penry *26*
Williams, W.T. 120
Winter, Sir John 21
Wood family, ironmasters
  Charles 42-43, *56*
  John 42-43, *56*
  William 31, 94
Wood, J.G., artist *24*, *25*, *28*
Woolf, Arthur 91
Wright & Jesson 43, *56*

If you are interested in purchasing other books published by Tempus, or in case you have difficulty finding any Tempus books in your local bookshop, you can also place orders directly through our website

www.tempus-publishing.com